THE   AMERICAN   LIBRARY

10, RUE DU GÉNÉRAL CAMOU
75007 PARIS

BOOKS BY DAVID RITCHIE

Spacewar   1982
The Ring of Fire   1981

SPACEWAR

SPACEWAR

DAVID RITCHIE

(New York **ATHENEUM** 1982)

Portions of this book appeared in different form
in the September 1, 1980, issue of *Inquiry* magazine.

All photographs are courtesy of the National Aeronautics and
Space Administration

Library of Congress Cataloging in Publication Data

Ritchie, David———
　　　　Spacewar.
　　Bibliography: p.
　　Includes index.
　1. Space warfare.　I. Title.
UG1530.R57　1982　　　358'.8　　　81-69153
　ISBN 0-689-11264-5　　　　　　　AACR2

Copyright © 1982 by David Ritchie
Published simultaneously in Canada by McClelland and Stewart Ltd.
Composition by American–Stratford Graphic Services, Inc.,
Brattleboro, Vermont
Manufactured by Fairfield Graphics, Fairfield, Pennsylvania
Designed by Mary Cregan
First Edition

TO BARBARA BOVA

Acknowledgments

This book began as an article for the September 1, 1980, issue of *Inquiry* magazine. Since then, so many persons have helped with the making of *Spacewar* that it would be impossible to list them all here. Chief among them are Barbara Bova, my literary agent; the editors at Atheneum Publishers; Glenn Garvin, editor of *Inquiry;* Ben Bova, editor of *Omni,* who read the original book proposal and made highly useful comments about it; Greg Miller, who provided material about the early proposals for spaceflight; and George Androvette, who probably heard more about this book, in its formative stages, than he ever wished to hear.

The libraries of Brandeis University, Northeastern University, and the Massachusetts Institute of Technology provided much valuable information from their periodical collections, and the faculty of the biology department at the University of Massachusetts in Boston were patient with me when it appeared I was spending more time on the book than on my duties there.

Contents

A Note on Measurements

Because this is a metric world, and the clumsy English system of weights and measures is gradually being phased out in the United States, all units given in this book are metric. Fortunately, conversions from English to metric, and vice versa, are simple. A meter is just over one yard; a kilogram is a bit over two pounds; a man of average height is about 1.75 meters tall; and so forth. For more accurate measurements, here are conversion factors:

One meter = 1.09 yards = 3.28 feet

One mile = 1.6 kilometers

One pound = 0.45 kilograms

One ounce = 28.4 grams

One quart = .95 liter

One ton (English or "short" ton) = 0.91 metric ton

Time measurements are of course the same in both systems.

The great advantage of the metric system is that it is arranged in powers of 10, so that computations are far easier than in the English system. There is no need to remember, for example, that a mile contains 5,280 feet when one measures in metrics; a kilometer is simply 1,000 meters.

We have no adequate idea of the predisposing power which an immense series of preparations for war has in actually begetting war.

WILLIAM GLADSTONE

SPACEWAR

ONE

"Bloodless War"

THE TIME : the not too distant future.

The place: approximately 150 kilometers above the earth.

At about twenty-five times the speed of sound, a United States reconnaissance satellite soared above the Pacific Ocean. Roughly the dimensions of a highway tank truck, the satellite was a supreme achievement of American aerospace technology, and was as far advanced over the early spy satellites of 1960 as the modern racing cars at Indianapolis were over Henry Ford's Model A.

This satellite, and others like it, formed an integral part of the U.S. defense network. What it reported back to earth might prevent wars, or determine the outcome of any that might break out.

An orbiting set of artificial eyes more sensitive than any ever evolved on earth by nature, the spacecraft was designed to keep watch on practically the whole inhabited area of the globe, from the equatorial seas to the remote defense and weather outposts above the polar circles. The satellite's cameras peered regularly at all the isolated corners of the world where preparations for war might be under way. It did not witness the fall of every sparrow, but it came close.

Inside the cylindrical hull, behind arrays of mirrors and lenses, was exquisitely sensitive film, capable of nearly infinite enlargement without serious loss of detail. This combination of superfilm and supercamera could pick out the lines on a soccer field in Poland, or highway direction markings in Odessa.

In theory, the satellite could "see" objects the width of a woman's hand. In practice, it was believed to do much better than that. Rumor had it that the spaceborne cameras could reproduce the front page of an issue of *Pravda* lying on a bench in Red Square right down to the photo captions. Employees at the plant that made the satellite had been heard to joke that only God had a more comprehensive view of the world than this space vehicle had.

Some of the stories were probably fantasy. In fact, no one outside a few labs, factories, and inner circles of government knew what the orbiting cameras could do. Their performance and all other specifications were secrets, except for a few details the government had seen fit to leak from time to time.

The government had been careful, however, to avoid telling the Soviets anything they might not know already about the satellite and its performance. Even the few tidbits of data thrown out to the press were years old, and it seemed likely that the state of the art had advanced greatly since those facts were current.

About all the average American (and many senators and congressmen as well) knew about the U.S. reconnaissance satellites was that their capabilities were phenomenal; that the Soviets presumably had their equivalent in orbit, watching us as we watched them; and that there was a wide variety of satellites circling the earth, monitoring everything from troop movements to military electronic communications. That latter job was handled by a special kind of satellite called a "ferret," which was said to have such sharp electronic ears that it could pick up walkie-talkie chatter between soldiers in the field.

Since the first military reconnaissance satellites had gone up in the early 1960s, they and their descendants had done as much as anything else to preserve the fragile peace between East and West. They insured that neither side could do anything significant, from a military standpoint, without the other side's knowledge.

To maintain their watch on the world, the American and Russian spy satellites had their orbits inclined steeply to the equator, so as to carry them over all the world's land area and most of its seas. Their field of view ranged from the icy Russian port of Murmansk in the north, with its giant submarine base, to the chilly Antarctic waters in the south.

Every day, more than 90 percent of the earth's surface passed under the glassy gaze of orbiting lenses. And every year, millions of images—some unimportant, some interesting, some matters of life and death—were focused on film and returned to earth for scrutiny. Some satellites transmitted their pictures to earth electronically. Others ejected special re-entry capsules full of film, which were snatched out of midair during the last few meters of their descent by aircraft equipped with special "lassos."

This particular U.S. satellite had photographs aboard that would interest analysts in Washington.

The Chinese were sending another division to the Soviet border. Was this movement a prelude to open hostilities, or just a precautionary measure?

There was good news from Eastern Europe. Things were quiet there for the moment. It looked as if the Soviets were undisturbed by political unrest in their European satellite countries. Or perhaps Moscow was too preoccupied with China's military buildup along the U.S.S.R.'s southern border to give Europe much attention.

That judgment would have to be left to humans, for the satellite was merely a sensing and recording device: a highly advanced one, to be sure, but still a machine. Even in the age of "smart" weapons and artificial intelligence, all the important decisions still rested with the most advanced computer, the human brain—which was why satellites like this one were needed in the first place. Their job was to give decision-makers the best information possible, a job they did so well that military and civilian officials on earth had come to rely on them for more and more activities.

Indeed, it was hard to imagine the U.S. and the Soviet

Union—particularly their national defense forces—without satellite technology. If either side lost the use of its satellites for even a few hours, the results might be catastrophic.

Noiselessly the satellite sailed over the ocean. Far below, clouds streamed out downwind in the lee of the Hawaiian Islands. To the north, a curved swath of white marked the advance of a cold front over the Aleutians. Near the equator, a milky spiral announced the birth of a tropical storm. And in the same vicinity, much smaller than the storm but still clearly visible to the satellite, was the foamy wake of a naval vessel—a Soviet aircraft carrier scrambling to get out of the typhoon's way. Click. Another snapshot for the folks back home.

A few moments earlier, a satellite had been launched from a site in the northwestern U.S.S.R. The launch was announced officially as just one more in the seemingly endless Cosmos series. The great white rocket booster climbed rapidly into space, driving its payload upward into an orbit just below that of the U.S. spy satellite, but identical in inclination.

As it neared the American satellite, the Russian satellite suddenly fired its rocket engine. The change in velocity nudged the Russian up into a higher orbit. Its new path took it within a hundred meters of the U.S. spacecraft.

Ninety meters away, the attack satellite exploded. A flying wall of shrapnel slammed into the U.S. satellite and riddled its hull.

In a few seconds, a technological marvel had been reduced to a few tons of scrap.

The attack was unheard, there being no air in space to

carry sound. But it was not unobserved. Detected by radar, the interception was followed at a special space-operations center buried deep beneath the Rocky Mountains. Just as satellites in orbit watched everything happening on earth, this nerve center kept track of everything happening in space. Every step of the incident registered on the screens there. The Russian satellite was seen approaching, putting on speed to overtake the American, and then blasting it into oblivion.

Of course, this could have been an accident. Space was becoming so cluttered with satellites and debris from launch vehicles that collisions between objects in orbit were almost commonplace. Only a few years earlier, such a chance encounter had demolished a U.S. communications satellite. So there was some possibility that this occurrence was just a fluke of probability. The peculiar pattern of the interception, however—especially the "pop-up" maneuver from a lower orbit to a higher one—made a deliberate assault seem all but certain.

A call went out to the White House, and from there another call to the Pentagon. Shortly afterward, a high-altitude jet fighter/interceptor took off from an Air Force base in New England. Flying almost straight up, it soon reached the edge of outer space, and leveled off.

From here the pilot had a spectacular view. Below was the blue earth, its curvature plainly visible from this height. The fuzzy cloak of atmosphere shaded gradually into the blue-black of interplanetary space. To the pilot's left hung the moon. It looked oddly small here, beyond the magnifying influence of air.

But the vista around him was not on the pilot's mind. On command from the ground, he touched a control at his right hand, and an electronic signal went out from the cockpit to a slender multi-stage rocket strapped under the airplane's right wing.

Instantly, the rocket dropped from its mounting. A split second later the engine ignited. The missile streaked out ahead of the aircraft, then curved upward into the darkness overhead, trailing a hyperbola of fire.

The first stage exhausted its fuel and dropped away, to tumble back toward the ground and incineration. The second stage cut in, and pushed the third and final stage into orbit before falling away like its predecessor.

The third stage did not ignite immediately. It was a small object, only about the size of an office wastebasket. It weighed about as much as a loaded suitcase. The rocket's final stage was loaded too, but with explosives.

Almost at orbital velocity, the little spacecraft sailed along. One polished side reflected the brilliant earth below; the other side, only space and stars. In the nose were special heat-seeking devices that could pick up the infrared emissions of spacecraft against the relatively cool backdrop of space. The sensors were "sniffing" the area as the satellite flew, searching for the target that American military intelligence had known would be nearby.

And there it was, a few kilometers ahead: a Soviet reconnaissance satellite, glowing brightly in IR wavelengths.

Now the engine fired. A slight course correction, and the U.S. interceptor missile was headed directly for the Russian satellite.

A hundred times more massive than its assailant, the Soviet spacecraft made no attempt to evade it. There was no escape, and the end was swift and violent.

9   "Bloodless War")

The missile slammed into its quarry's midsection at a relative velocity of five kilometers per second. Bits of metal flew in all directions as the impact buckled and shattered the Russian hull. The explosion ruptured the satellite's propellant tanks, and from them a cloud of pale vapor spread out through space, glowing like morning mist in the sunlight.

Far below, at a site in the American Midwest near the edge of a wheat field, a camera with an aperture three meters wide watched the demise of the Soviet spacecraft. The camera could detect something the size of a soccer ball 40,000 kilometers up, and had no trouble picking out the images of the Russian spacecraft and its tiny assassin.

On impact, the U.S. interceptor could be seen to vanish in the puff of vapor from the Russian's hull. Then the spreading fog enveloped them both. When it finally dispersed, the wreckage of the Soviet craft could be seen spinning through space, smashed and crumpled like a beer can.

More than half a billion Russians and Americans went about their daily business as usual, unaware that a battle had been taking place above their heads.

This little melodrama is fiction, but something like it may take place in the next decade or two, if it has not occurred already.

In recent years, "duels" in outer space have graduated from the realm of pulp fiction—the Buck Rogers and Flash Gordon genre—to grim possibility. When and if they occur (and some arms control experts believe that "if" is merely wishful thinking), battles like this one will represent the culmination of a set of trends and developments that began some forty years ago.

Since the end of World War II, the technology of warfare has been progressing at an awesome rate. In place of the propeller-driven fighter aircraft of 1945, the world's air forces now have aircraft capable of flying at two or three times the speed of sound, and equipped with electronic combat aids undreamed of even twenty years ago. Jimmy Doolittle's men dropped bombs whose accuracy depended mostly on the bombardier's own judgment; today bombs are made smart with computers and laser targeting systems, so that the bombs have, theoretically, a better than 90 percent chance of hitting home.*

The relatively crude tanks of the North African campaign have been replaced by electronically advanced land monsters that can "see" in pitch darkness and will enable their crews to survive a nearby nuclear explosion. And the diesel-powered U-boats have evolved into nuclear-driven submarines almost as populous as small cities, each sub loaded with enough atomic firepower to wipe out entire nations.

Warfare on land, at sea, and in the air has changed in ways that would have astonished the generals and admirals of Franklin Delano Roosevelt's day, had they been able to see a few years into the future. And because each

* In World War II, the chance was nearer 1 percent that a bomb would strike its designated target, unless that target was the size of a factory. Still, that effectiveness was superb compared with that of some other weapons. *One hundred thousand* rifle and machine-gun bullets were fired in World War II for every one that hit a soldier.

advance spurred still more advances in military hardware
and tactics, the technology of armed conflict is exploding
—both literally and figuratively—as never before.*

So it is no surprise that in the past three decades science
has introduced warfare to the last remaining arena: outer
space.

Consider the following information:

The Soviet Union has been testing "hunter-killer" satellites since the late 1960s that are said to be capable of
knocking U.S. reconnaissance and communications satellites out of the sky at will. Russian anti-satellite (ASAT)
weapons are thought to have scored more than a dozen
successful "kills" on targets in orbit. U.S. Department of
Defense (DOD) officials are concerned that the Russians
may have the capability to intercept American spacecraft
within a single orbit of the attack vehicle. As *Aviation
Week and Space Technology* magazine (hereafter referred to as *Aviation Week*) has stated in an article
about Russian ASAT weapons, an attack carried out within
the first orbit of the "killer" satellite might not give the
United States enough time to detect that one of its spacecraft was under assault, and respond accordingly. To
avoid being caught by surprise, the U.S. has been forced
to identify "threat windows," or times when our satellites
may be in maximum danger from Russian ASAT weapons.

---

* Military leaders have not always embraced new technologies
as eagerly as they do today. Marshal Foch, looking at the flimsy
aircraft of the early 1900s, dismissed the airplane as a "noisy toy"
and said it would never be practical for combat. And Foch's
countryman Napoleon Bonaparte once declined to sponsor the
building of a French military submarine because he thought it
"ungentlemanly" to attack from beneath the waves.

Russia is said to be developing laser ASAT weapons. In 1975 there was a news report, denied by the Pentagon, that the Soviets had used a powerful laser beam, fired from the ground, to "blind" two American reconnaissance satellites. Not to be outdone, DOD so far has spent more than a billion dollars on laser weapons that might be used one day against enemy missiles and satellites. Some U.S. plans for laser weapons are grandiose, to say the least. One proposed system would use a mirror four meters wide —roughly twice the height of a grown man—to focus a laser beam strong enough to "fry" its targets. The laser beam would burn through the hull of an incoming missile and destabilize it, or would destroy the antennae and other vulnerable external parts of satellites.

If public statements by DOD officials are any guide, then the United States may soon have a whole new branch of the armed services—a Space Force, in effect—to handle combat operations in orbit. The Air Force already has a separate Space Division to coordinate all its efforts in space, and a man-in-space capability for DOD is likely in the near future. The Air Force has established a manned spaceflight support group at the Johnson Space Center in Texas, and this group may form the nucleus of a future Space Force.

Among the objects tracked by NORAD (North American Air Defense Command) are the 20-ton Soviet Salyut space stations. Salyut 6, presently in orbit, is thought to be the starting point for a much larger Soviet space platform, presumably one with military applications. The Salyuts, some of which have been manned, and others unmanned, are roughly comparable to our Skylab.

Though Salyut is nominally a peaceful project, it is known to be an important part of the Soviet military space effort, and at least one Salyut crew has been forced to fill in for a malfunctioning spy satellite.

All Soviet space activities are under military direction, and before long the same may be true of the American space program. Funding for American military space projects has surpassed the budget for the civilian space effort; and while the budget for the National Aeronautics and Space Administration (NASA), the civilian space agency, keeps shrinking, DOD's space expenditures keep rising. Indeed, DOD is eclipsing NASA so quickly that there is serious talk in Washington of merging NASA with the Defense Department—thus putting an end to the U.S. non-military space program.

The militarization of space will affect—and is affecting now—the lives of virtually everyone on earth. The way it came about is a story of politics and technology, and how a few inventions combined—in ways their creators could never have intended—to change our world irrevocably.

TWO

"How and Why Was This Allowed to Happen?"

T H E   S T U D Y   of history is full of ifs. *If* Robert E. Lee had accepted President Abraham Lincoln's offer of command of the Union armies; *if* Mohammed had remained a businessman; *if* Napoleon had been spared an attack of hemorrhoids just before the battle of Waterloo; *if* F.D.R. had lived to complete his fourth term as President—if any or all these things had happened, then the world we live in might be a markedly different place.

And if the history of rocket technology had followed a

slightly different course in the third and fourth decades of the twentieth century, then Europe might be united today, under Nazi rule; a space shuttle might be making routine trips into orbit, but with a swastika on its tail instead of the Stars and Stripes; and Americans might still be mourning the destruction of New York City in 1945 by a German spaceship armed with an atomic bomb.

Pure fantasy? Not at all. Long before the U.S. and Soviet space programs were more than pipe dreams, the rocket designers of the Third Reich had plans for manned military spaceships similar in many ways to the modern American space shuttle.

United States and Soviet military space vehicles now being built and flown are descendants of those early German craft, in form as well as purpose. And if the Nazis had had a bit more time to work on their military rockets, and had been able to join them with another weapon the Germans came within a few months of building, the war might have ended in victory for Hitler.

Ironically, one of the men who made the evolution of military spaceships possible was a peaceful, balding schoolteacher from central Massachusetts. His name was Robert Hutchings Goddard, and his pioneering work in rocket technology taught the Germans almost everything they knew.

Goddard was born in 1882, in the Roxbury neighborhood of Boston, which was then a pleasant suburb of white frame homes and tree-lined streets, and Goddard appears to have had a conventional, happy boyhood there. He became infatuated with the idea of space travel at an early age, after reading fantastic novels of adventure on other worlds. One of his favorite books was

H. G. Wells's *The War of the Worlds*. He was also impressed by a story called *Edison's Conquest of Mars*, by a writer named Garrett P. Serviss.

Mars was a common setting for adventure stories in Goddard's youth, because New England astronomer Percival Lowell had popularized the vision of Mars as a dying world inhabited by a technological species like ourselves. During periods of good visibility, Lowell, seated at his telescope, thought he could just barely make out long parallel lines on the surface of Mars. A man of imagination, Lowell speculated that these lines (which he assumed were material objects of some kind) must be canals that the Martians had built to channel precious water wherever it was needed on their dry and moribund world. It was an appealing thought, and soon the press was full of stories and artists' visualizations of the great Martian canals and cities. And while Lowell's canals turned out to be totally illusory—space probes reaching the planet several decades later revealed nothing like them—the idea of an inhabited Mars was a godsend to fiction.

No author got more mileage out of Lowell's vision than Edgar Rice Burroughs, creator of Tarzan. Burroughs wrote a string of novels set on Mars, which he populated with virile heroes, seductive damsels, and all manner of monsters. These novels fascinated young Robert Goddard and made him dream of reaching the red planet.

The novels were flawed, however, because Burroughs had no plausible means of hauling his protagonists from one world to another. In his novel *John Carter of Mars,* the hero simply closes his eyes and finds himself transported to Mars by some inscrutable cosmic force. This was clearly an unreliable means of space travel, and Goddard decided that if humans ever were to reach other worlds, the only way to do so would be by rocket.

Rocket research was Goddard's hobby when he was an undergraduate at Worcester Polytechnic, in Worcester, Massachusetts. He went on to do graduate work at Clark University, also in Worcester, and received his Ph.D. in physics in 1911, at the age of twenty-eight. Shortly thereafter he took a teaching position at Clark and, as a supplement to his lectures on math and physics, discussed ways of penetrating outer space. His students listened as Goddard spoke in the arcane language of rocket science: thrust, drag, escape velocities. Most likely no one in the classroom, Goddard included, had any idea then how the material in these talks would reshape the world in the years ahead.

Goddard planned to start with small rockets and gradually work his way up to large ones capable of lifting payloads out of the atmosphere. Fortunately—since he could afford very few experiments on the salary of a college teacher—the Smithsonian Institution was impressed by his monograph "A Method of Reaching Extreme Altitudes," and in 1916 granted him $5,000 for rocket research.

During World War I, Goddard worked for the Army, developing a recoilless rocket that a soldier could fire against enemy troops. Goddard succeeded, but too late for his invention to be used in the war. In World War II, however, Goddard's invention was refined slightly and became one of the most effective anti-tank weapons in history: the bazooka. (Another of his ideas, an echo-location system for detecting U-boats under water, became what we call sonar.)

(SPACEWAR 18

After World War I, Goddard went back to work on rockets for peaceful purposes. He was aided by the Guggenheim Foundation and by his friend Colonel Charles Lindbergh, who persuaded the foundation to back his research.

At about the same time, Goddard was worried over news he had been receiving from Germany, where scientists were at work on liquid-fuel rocket designs like his. Goddard distrusted the Germans, because he thought they had no interest in rockets except as weapons; and, as a former designer of such weapons, he knew perhaps better than anyone else what rocket technology might do in the hands of the German armed forces.

He was disturbed by letters such as one that reached him in 1922 from a student at Heidelberg named Hermann Oberth. In shaky English, Oberth explained his interest in Goddard's research and asked for copies of some of his writings.

Despite misgivings, Goddard decided to be courteous and sent Oberth a copy of "A Method of Reaching Extreme Altitudes." Later, Goddard regretted his generosity, for Oberth sent him a copy of his 1922 book *Die Rakete zu den Planetenräumen* (*The Rocket Into Interplanetary Space*), in which he carried his speculations much farther than Goddard had. Instead of mere rocket shots into the upper atmosphere or to the moon, Oberth was proposing voyages to other planets, and he backed up his recommendations with solid mathematics.

It was the kind of book Goddard understood all too well, and when he saw how far the Germans had advanced in rocketry he was alarmed—especially since he felt he had probably contributed to their progress by sending his paper to Oberth. Goddard corresponded briefly with Oberth thereafter, but wished to have as little contact with "that German" as possible.*

On March 16, 1926, Goddard succeeded for the first time in launching a liquid-fuel rocket, on his Aunt Effie's farm in Auburn, Massachusetts. The rocket was fueled by gasoline and liquid oxygen. It stood about twice as tall as Goddard himself and looked like a cross between a firecracker and a coat hanger. At 2:30 P.M., Goddard stood back and touched off the ignition. The rocket rose from its launching cradle, flew 59 meters horizontally, and reached an altitude of 13 meters. Flight time was 2.5 seconds. The rocket landed in Aunt Effie's cabbage patch.

Goddard's experiments continued until 1929, when, after an especially noisy test, the Commonwealth of Massachusetts declared his rockets a public nuisance and told him he could never launch them again on any land under state jurisdiction. So he moved his hardware to a remote part of an Army artillery range at Camp Devens (now Fort Devens), Massachusetts. The range was federal property, so Goddard was free to continue his work without interference from state authorities. Soon afterward, he acquired a better test site in Roswell, New Mexico, and moved there with his wife and a team of assistants that included his brother-in-law.

Remote though Roswell was, Goddard was not isolated from news from abroad. The British, French, Russians, and Japanese were all taking a keen interest in rocketry,

* In fact, Oberth was born in Transylvania and never considered himself a German at all. He applied for German citizenship only when the Nazis convinced him that the alternative was a concentration camp.

thanks in part to the inspiration Goddard had provided. He was still worried, however, about what he considered the sinister progress of the Germans.

By 1930, Germany's Verein für Raumschiffahrt (Society for Space Travel) had several hundred members and by 1931 had succeeded in launching a liquid-fuel rocket. American visitors to Germany returned with impressive eyewitness accounts of rocket tests. Moreover, Goddard must certainly have shuddered if he saw the 1929 German motion picture *Die Frau im Mond* (*The Woman on the Moon*), which featured a model moon rocket similar in many respects to the Apollo spacecraft that reached the moon four decades later.*

Goddard's concern about the Germans increased during the 1930s. Letters he received from German rocket specialists revealed how far along they were in developing liquid-fuel rockets. They were even offering *him* ideas. In 1935, Goddard received a communication from one Eugen Sänger, who would later have a profound influence on the military uses of space, suggesting a technique for pumping liquids into a rocket's combustion chamber.

Then, in 1939, inquiries to Goddard from German scientists ceased abruptly. This was an almost certain sign that something important was happening in German rocketry—probably something with a military bent. That same year, Lindbergh all but confirmed the suspicion for Goddard after a visit to Germany; the Nazis showed off their newly developed air power, but when Lindbergh

* Director Fritz Lang originated the now familiar countdown before launch for this movie. He called it one of his "damned dramatic touches," never suspecting it would one day become an actual preflight ritual.

asked about German rockets, his hosts were not the least bit forthcoming. What was going on in the Third Reich's rocket labs?

The answer could be found on Germany's northern coastline, at a place called Peenemünde. Located where the Peene River meets the sea (Peenemünde is pronounced "pay-neh-*myoon*-duh" and means "mouth of the Peene"), Peenemünde was the Nazis' secret rocket development center. A few miles north of the Bay of Stettin, on a barrier island between the bay and the Baltic Sea, Peenemünde was admirably situated for rocket research and production. The rural location made espionage difficult, and the nearby ocean provided a measure of safety: rockets gone haywire would simply fall into the water, where they would harm no one, and where their wreckage could not be gathered by spies and sent to Allied intelligence for study. Peenemünde also had a third advantage—proximity to the British Isles, the principal target for the weapons built there.

Peenemünde was actually two installations working together to build the forerunners of our modern intercontinental ballistic missiles (ICBMs) and cruise missiles, the famous V-weapons. The "V" designation was the idea of Nazi propaganda minister Joseph Goebbels, and stood for "vengeance."

One section of the rocket base was called Peenemünde East and was run by the Wehrmacht (German Army), while the Luftwaffe (Air Force) had charge of Peenemünde West. The eastern branch of Peenemünde was the rocket development area. Altogether, some 20,000 persons were employed at Peenemünde, including several who would later play important roles in the American military space effort.

(SPACEWAR 22

The earliest and simplest of the V-weapons was the Luftwaffe's V-1, a pilotless flying bomb about six meters in wingspan, with an average range of about 250 kilometers—just enough to carry the bombs into southern England. The V-1 looked much like a conventional airplane, with one exception. Mounted at the tail, atop the vertical stabilizer, was a strange-looking engine called the *Strahlrohr,* or "jet pipe." It consisted of a tube with a set of metal shutters at the forward end. Air pressure forced open these shutters when the vehicle was in flight, and air entered the tube to be mixed with a fine fuel spray. When the mixture was ignited, combustion products rushed out the rear as exhaust. The explosion created a partial vacuum inside the tube, and air pressure from outside forced open the shutters again, starting the whole cycle over. There were several cycles per second, each cycle producing an impulse of thrust. The resultant noise was compared to the amplified hum of a bee, and so the English, for whom the V-1 was intended, dubbed it the "buzz bomb" or "doodlebug."

Just before impact, as its fuel ran out, the doodlebug would fall silent. There was then a short pause as the bomb dropped out of the sky, followed by the explosion of the one-ton warhead.

British writer George Orwell, author of *1984,* became familiar with the ominous silence that preceded a buzz bomb's landing. In a wartime essay, he noted that the brief delay gave one just enough time to hope the bomb hit someone else—and to reflect on the "bottomless selfishness" of the human species.

From mid-June to September of 1944, an average of a hundred V-1s took off for England every day from launch sites along the northern coasts of Germany and occupied Europe. Some days, as many as 295 were launched in twenty-four hours, or about one every five minutes.

The V-1 was not a terribly fearsome weapon, however. It flew low, and was so slow that fighter aircraft could shoot it down. Doodlebugs were easy prey for anti-aircraft guns and could be ensnared in the dangling cables of barrage balloons. But these countermeasures cost money, and this was in fact the principal achievement of the V-1s: to divert much-needed funds to doodlebug defense. A 1944 British Air Ministry study concluded that for every dollar the Germans spent on their flying bombs, the British had to spend four dollars to neutralize them.

The V-1, it should be remembered, was not a true rocket. It was more a jet, because it required oxygen from the atmosphere to run the *Strahlrohr*. The first real rocket weapon to descend on Britain was the buzz bomb's successor, the V-2, the first long-range liquid-fuel missile and one of the most important military inventions in history.

Originally known as the A-4 (the "A" stood for *Amalgam*, the German equivalent of "prototype"), the V-2 was the first modern ballistic missile and served as a model for the American and Russian liquid-fuel launch vehicles that followed it. About 12 meters long from the tip of its warhead to the trailing edge of its tail fins, the V-2 was propelled by liquid oxygen and a 75 percent solution of alcohol. A V-2 could deliver one ton of amatol high explosive to a target 250 kilometers away in five minutes. That was no mean feat for the 1940s.

One September 8, 1944, a V-2 took off from a launch site in northern Holland, arched toward the northwest at an angle of about 45 degrees, and rose to an altitude of

100 kilometers. Then the missile, its fuel spent, fell back along a ballistic trajectory toward Britain.* The V-2 landed in southern England, the first of many that would bombard targets in Britain, France, Holland, Belgium, Poland, and even, in the last desperate days of the war, Germany itself.

This was a weapon entirely different from the slow and clumsy V-1. For one thing, there was no adequate defense against a V-2 attack. As noted, the doodlebug could be shot down or caught by balloons, but such measures were ineffective against a V-2. The huge rocket hurtled down from the edge of space so fast that one could barely see it at all, much less bring guns to bear on it. Also—and this was perhaps the most terrifying thing about the V-2—it gave no warning of its approach. The V-1 had announced its coming with its peculiar engine noise; but V-2s, by contrast, descended faster than sound, and so there was no noise to alert potential victims on the ground. One simply never knew when a giant Nazi missile with a ton of high explosive in its nose might be falling toward the roof of one's home. A standing house disappeared in an instant in the blast from a V-2 warhead. This awful uncertainty probably explains why the British coined no nickname such as "doodlebug" for the V-2: the thing was just too horrible. Britons thus became the first people to live in constant fear of descending ballistic missiles—a fear that Americans and Russians would learn some years later.

Among the rocket experts at Peenemünde was a young

* "Ballistic," as applied to missiles and re-entry vehicles, means the object hurtles along under its own momentum, with no power to correct its trajectory.

Silesian named Wernher von Braun. He had been tapped for military work by the Wehrmacht while experimenting with rockets outside Berlin in the 1930s. Along with Willy Ley (who later emigrated to the United States and became a well-known science writer) and other members of the old Verein für Raumschiffahrt, von Braun had long been fascinated with the concept of manned spaceflight— though preferably for peaceful rather than belligerent purposes.

Until the development of the V-2, there had been no rocket powerful enough to carry a pilot, unless one counted a few attempts to build small rocket-powered fighter aircraft.* What the Peenemünde team had in mind, however, was something more formidable than a small, short-range fighter plane. They were thinking of a manned, rocket-propelled bomber that would take off like a V-2, soar to the edge of space, drop bombs on a target thousands of miles away, and finally glide down to a controlled landing. So the Germans looked for ways to convert the V-2 into a piloted spaceship.

The first result was the A-4B, irreverently nicknamed the "bastard A-4." Little more than a standard V-2 with large swept-back wings bolted onto its body (the design had been tested in wind tunnels and seemed suitable for hypersonic speeds), the A-4B was ready for flight tests in January, 1945. The first test was a failure, but the second

---

* One of these, a midget aircraft called the Komet, was unleashed against Allied bombers in the final weeks of the war. Carried aloft by a Heinkel bomber and released, the tiny interceptor moved so fast that the bombers' gun crews could not follow it fast enough to score any hits. The Komet knocked them out of the sky almost before the Allied gunners could fire a shot.

A-4B reached an altitude of 125 kilometers and a velocity of 6,500 kilometers per hour. The A-4B thus became the first winged missile ever to break the sound barrier, and Germany's space-bomber effort was under way.

The next step was a piloted A-4B with tricycle landing gear. Designated the A-9, it was 14 meters long, had a takeoff weight of 16 metric tons, and could deliver a ton of bombs to a target 5,000 kilometers away—roughly the mileage between New York City and San Francisco. That range put all the major cities of Europe within the A-9's reach.

An A-9 mission would have gone something like this. The bomber would take off from Germany and fly to a maximum altitude of 165 kilometers. Then the pilot would put the craft in re-entry position, with the nose raised slightly, and start gliding down through the atmosphere on his bombing run. After releasing bombs over the designated target, he would bring the A-9 down to a landing at a conventional airstrip, or bail out if there was no airfield nearby in friendly territory. An A-9 mission would take only seventeen minutes from launch to landing.

As soon as plans were finished, several prototype A-9s were ordered built. But the A-9 was barely off the drawing board before the Germans were looking for ways to extend its range across the Atlantic to America.

To carry the A-9 over the ocean, the Germans conceived the A-10, a liquid-fuel rocket similar in many ways to the U.S. Atlas booster of the 1960s. The A-10 was 20 meters long and slightly more than 4 meters in diameter, and could deliver 200 tons of thrust.

In a little less than a minute, the booster would lift the

A-9 to an altitude of 24 kilometers and accelerate it to a velocity of 1.2 kilometers per second. Then the A-10, its fuel spent, would be cast away and fall back to earth, while the A-9's pilot cut in the rocket plane's engines and continued the flight. At the start of its bombing run, the A-9 would be 160 kilometers above the ocean and traveling three kilometers per second.

Clearly, an A-9/A-10 mission would have been an expensive undertaking. So, rather than use a new A-10 on every flight, the thrifty Germans planned to recover the boosters and use them repeatedly. When the A-10 ran out of fuel and was jettisoned, air brakes would unfold from the booster's body and slow its fall through the atmosphere. A few kilometers above the ground, the A-10 would deploy a parachute and drift down gently to a soft landing. This is the method NASA uses to recover the solid-fuel rocket boosters from the U.S. space shuttle, and is another sign of how advanced German rocket design was in 1945.

Yet the A-9 and A-10, advanced as they were, look primitive when compared to another German project of this time: the "antipodal space bomber." This remarkable spacecraft, which turned out to be a powerful influence on later designs, was conceived by the famous designer team of Dr. Eugen Sänger and Dr. Irene Bredt. In many ways, the Sänger-Bredt bomber was a smaller version of the modern NASA shuttle, adapted for dropping bombs.

A low-wing craft 30 meters in length, with a wingspan of 15.2 meters, the antipodal bomber was about the size of one of today's small jetliners. The fuselage housed two large rocket engines laid side by side. At the tail was a horizontal stabilizer with two stubby vertical stabilizers at the tips. On either side of the bomb bay, midway up the fuselage, were two absurdly small seeming-wings, only a little bigger than the stabilizers. The pilot sat in the nose of the bomber, just ahead of the liquid oxygen tanks. The Sänger-Bredt bomber had a flat belly for increased lift on re-entry, just as the U.S. shuttle has.

This strange-looking craft was designed to carry a ton of high explosive—the same payload as the V-1 and V-2 —and to take off from Europe for attacks on North America. Rather than being lifted into space by a liquid-fuel booster that would then be returned to the ground by parachute, the antipodal space bomber would be mounted on a cradle and hurled down a track three kilometers long, by a battery of rockets with 250 tons of thrust. In only 11 seconds, the rockets would consume 50 tons of fuel and bring the spaceship—still on the ground—to a velocity .5 kilometer per second, or just over 1.3 times the speed of sound. Bystanders would hear a sonic boom before the bomber had even left its launching track!

Near the end of the track, the pilot would pull back on the stick and bring the bomber into a steep unpowered climb. Then he would ignite the spaceship's own rocket engines at 10 kilometers altitude and climb out of the atmosphere at an angle of 30 degrees, still accelerating. Fuel would run out at 145 kilometers altitude, and the bomber would then coast along to its maximum altitude of 250 kilometers (about that of present-day photoreconnaissance satellites) before starting the attack.

The pilot would ease the ship back into the atmosphere, nose up and belly forward, at the start of a bomb run. Using the lift created by the flat belly—Bredt had run a mathematical analysis of the bomber's configuration and found that the flat underside would make the bomber "skip" across the top of the atmosphere, as a rock skips across the surface of a pond—the pilot would dip briefly into the atmosphere from time to time to slow down the bomber, then soar back into space to prevent air friction from heating the hull to a perilously high temperature.

On the next to last dip into the upper air, the pilot would release his bombs over the target. (Actually, he would have released them quite some distance away, because the terrific velocity of the spaceship would carry them many kilometers after they were dropped.) The lightened bomber would then weigh only ten tons and would bound back upward into space, to continue coasting around the world. At last the spaceship, robbed of much of its momentum by friction with the atmosphere, would glide down to a landing in Japanese- or German-occupied territory.

Although a prototype engine for the Sänger-Bredt bomber was reportedly built and tested, the bomber itself never flew. Neither, for that matter, did the A-9 nor the A-10. The war was over before either spaceship ever came close to a test flight. If they had flown, the German military spacecraft would have been the most costly and inefficient warplanes ever built. A single A-9 mission against Washington or New York would have cost hundreds of thousands, if not millions of marks, and would have delivered only a paltry ton of high explosive to the target: enough to knock out a few buildings, but no more. Such a small result would hardly have justified the expense.

What if the Nazis had been able, however, to join their military spacecraft with another weapon they came within a few months of completing: the atomic bomb? Armed with nuclear weapons, the Nazi space bombers would have been formidable machines indeed. A single antipodal space bomber with an A-bomb in the hold might have destroyed Washington, Boston, or New York—or London or Moscow—and thus brought the Allies to their knees. What anti-aircraft gun could touch a warplane that flew more than a hundred kilometers high? And how long could the Allies have held out against invincible spaceships armed with city-leveling bombs? Two nuclear bombs demoralized Japan to the point of surrender. Atomic bombing raids from space by Nazi spacecraft would probably have had the same effect on the U.S. and its allies.

Shortly after the A-4B was tested, however, the Germans at Peenemünde could see that the Third Reich was finished, and that their "miracle weapons," which Hitler hoped would turn the tide of war against the Allies, could not save Germany from collapse. The only important question was: to whom would the German rocket experts surrender?

The Russians? No, German scientists, even with highly sophisticated weapons to offer, could expect little sympathy there. The memories of Stalingrad, of Kursk, and of the breaking of the Hitler-Stalin Pact were still fresh in the Russians' minds.

The British and Americans seemed the best prospect, and so one morning Wernher von Braun's brother Magnus, who spoke English well (and was "the most expendable" of the Peenemünde group, he recalled later), rode his bicycle to the American lines a few miles away and said the engineers at Peenemünde wanted to surrender.

That's fine, the Americans replied; tell them to come on over, and bring their rockets with them. Thus the United States acquired the cream of Germany's rocket designers, along with their hardware and accumulated knowledge, in just a few hours.

Stalin was enraged when he heard of the Americans' luck at Peenemünde. "Intolerable!" he roared, and his fury only increased when he learned that the Americans had smuggled a whole warehouse of assembled V-2s and spare parts right out from under the Russians' noses. To General I. A. Serov, of the Soviet Military Administration in Germany, Stalin wrote: "How and why was this allowed to happen?" Serov would later hear much more from Stalin on the subject of Germany's military rockets, and would join in a cloak-and-dagger operation to try to recoup the Russians' loss to the United States.

While Stalin fumed, the Germans were talking to the British and Americans, and astonished the latter by telling them how much the German war rockets owed to Robert Goddard's research. One German officer paid tribute to Goddard in these words: "Professor Goddard was one of the outstanding rocket pioneers in his country. We could not understand that a man of his genius did not get sufficient support of his government in time." That same year, shortly before his death, Goddard himself had a chance to see how his work had contributed to the Germans' achievements at Peenemünde when he inspected a captured V-2. In all but a few details, it was identical to some of his larger rocket designs.

Goddard's flimsy-looking rocket, fired one day on his Aunt Effie's farm less than two decades earlier, had led in that short time to the V-2. In the next twenty years, the

U.S. and Soviet armed forces revived plans for manned space bombers, and conceived other new weapons for use in space that made the V-2 look as crude, by comparison, as Goddard's early bazooka.

THREE

"We Have to Have Them Sooner Than Our Potential Enemy"

THE SUPERSTAR among the Germans was, of course, von Braun. Handsome and good-humored, with a broad toothy grin, von Braun never lost his Teutonic accent, but referred to it as his "Southern drawl" after the Army sent him to Huntsville, Alabama, to work on the military rocket program at Redstone Arsenal.

Von Braun, who had an uncanny understanding of how the American media worked, became a celebrity in the U.S.; almost as soon as he set foot on American soil, he was courting the press. His first major media coup came in 1952, when he joined Cornelius Ryan, author of *The Longest Day* and *A Bridge Too Far*, and illustrator Chesley Bonestell (who had done the artwork for the opening sequences of George Pal's Oscar-winning film version of *The War of the Worlds*) to produce a series of articles for *Collier's* magazine. Under the title "Man Will Conquer Space Soon," von Braun and Ryan outlined a plan to put astronauts into space.

The first spacemen would be preceded into orbit by rhesus monkeys, or other experimental animals, to gather vital information on how passengers in a spacecraft would react to such phenomena as the terrific gravity, or g-forces, of a launch and the eerie state of weightlessness. To spare them the torture of re-entry, when g-forces would build up to possibly lethal levels and the whole spaceship might be incinerated from air friction, the test animals would receive a lethal dose of toxic gas just before their craft plunged back into the atmosphere.

Next, von Braun proposed sending up manned spacecraft that sounded like enlarged versions of the A-9/A-10 combination envisioned at Peenemünde (without the bomb bay, of course). The booster would be recovered by parachute, and the spaceship itself would have wings for a controlled descent. Eventually, von Braun suggested, space stations could be assembled in orbit for use as military outposts; high-resolution telescopes on board the stations could be used to spy on Communist countries; and from the stations guided missiles armed with nuclear warheads could be launched against hostile targets on earth, if the need arose. (About a decade after von Braun proposed such orbiting fortresses, the Air Force took that concept and turned it into the Manned Orbiting Laboratory, or MOL, which will be the subject of a later chapter.)

Not everyone embraced von Braun's ideas. But von Braun's siren song of space was hard to resist, for he was a charismatic man with evangelical zeal, and he took every opportunity to put the case for space before the public.

Some of the projects von Braun favored were grandiose, to say the least. He advocated a plan to build a nuclear-powered spaceship thirty stories tall, named Orion.* Rather than rockets, Orion would use atomic explosions to lift itself into orbit. Small nuclear bombs, "shaped" so as to concentrate most of their blast upward against a metal "pusher plate" at the spaceship's stern, would be dropped out the rear of Orion and set off a couple of hundred feet from the plate. Formed like a giant bullet, Orion would rise into space on a pillar of nuclear fire, and would have so much energy to spend that there would be no need to minimize weight in its design. Orion could have carried practically anything that would fit into its hold—even heavy machinery. (A conventional chemical-fueled rocket, by contrast, must save weight wherever possible, because it takes so much propellant to put a payload in orbit.)

Nothing ever came of Orion. A model was built and test-flown, using small explosive charges for propulsion, and the giant spaceship got the backing of some influential Defense Department officials. The U.S.-Soviet nuclear test ban treaty put an end to atomic blasts in the atmosphere, however, and so there was no way to get Orion off the ground without breaking the pact.

* Not to be confused with a much later NASA project by the same name, designed to detect planets orbiting distant stars.

Such schemes made von Braun excellent copy for the press, and in the light of his fame some other brilliant alumni of Peenemünde were overlooked by the media. One of those scientists was Dr. Walter Dornberger, von Braun's boss at the Baltic rocket research base.

An avuncular-looking man with penetrating eyes, Dornberger started his career in rocketry in 1930, when the Wehrmacht assigned him to the ballistics branch of the Board of Ordnance, with the job of developing solid-fuel rockets. When the Peenemünde base was set up six years later, Dornberger was put at its head and supervised the development of the V-weapons. He came to the United States in 1947 and gave his expertise to the U.S. Air Force guided missile effort. In 1950, he joined the Bell Aerospace Company as a consultant. Ten years later, he had worked his way up to a position as chief scientist and vice-president of the company.

Dornberger had never forgotten the dream of manned military spacecraft like the A-9 and the Sänger-Bredt antipodal space bomber. Now, only a few years after the fall of Germany, rocket technology had advanced to the point where such spaceships were practical, and shortly after his arrival at Bell, Dornberger moved to make that vision a reality.

Dornberger talked Bell into conducting a top-secret study of a space-bomber design. Officially called Project Bomi, it was also known as "the Dornberger project." Bomi was a two-stage craft, rather like the old A-9/A10 combination. In Bomi, however, *both* stages were manned. A big delta-winged first stage would carry a second, smaller winged vehicle to the edge of outer space, where the two stages would part company. The upper stage would proceed into orbit under its own power, while the lower stage glided back to earth. When its mission in space was done, the upper stage of Bomi would return to the ground in like manner.

Bomi was a totally reusable spacecraft, and it was designed to subject its pilot and passengers (there might be occasions when it would haul persons rather than hardware) to no more than 3.5 times the normal force of gravity. Even an elderly person in decent health could tolerate such mild g-loading for short periods. At a peak velocity of about 5,000 kilometers per hour, Bomi could have crossed the United States from California to New York City in an hour and fifteen minutes.*

Bomi, like Orion, never flew. But the Air Force took an interest in Bomi, and by the late 1950s various similar space vehicles were under study. There was 118P, a high-flying reconnaissance craft, and Brass Bell, another spy ship. There was also a project code-named ROBO—short for "rocket bomber." Given the money and the official backing, the United States probably could have had something like the Sänger-Bredt antipodal bomber flying before 1960. American aerospace engineers were strongly attracted to the idea of a piloted, winged re-entry vehicle. The design had been worked out in detail by several different U.S. aerospace firms, as well as by the Germans, and about all that remained to be done was to scale up a liquid-fuel booster to the size needed to orbit a spaceship.

Under the Eisenhower administration, however, backing was hard to obtain for a military man-in-space effort. Eisenhower himself was wary of engrossing the military-

* Dornberger invited Eugen Sänger to come to the United States and work on Bomi, but Sänger refused.

industrial complex, and his attitude filtered down through the executive branch and permeated the U.S. government as a whole. At the same time, the administration was lukewarm at best toward space exploration in general. In fact, the federal government, through the armed services, tried to discourage the launching of earth satellites. The Air Force was forbidden even to use the word "space" in official remarks. And while the Air Force was given $10 million for space projects, it was told to steer clear of "system development"—a catchall term that included manned spacecraft like Bomi.

Then came Sputnik.

On October 4, 1957, the first artificial earth satellite rose into space atop a Russian rocket. Russia's initial satellite was crude by modern standards, but it had a shattering effect on American morale. For years, Americans had been told—and had believed—that the tyrannical Red rulers and their unwashed peasant subjects were incapable of matching the mighty West in technology. And now this *thing* was beeping across the sky, in plain view of the entire world, as incontrovertible proof that Moscow could indeed give the U.S. a run for its money.

Official comment from the United States was mild at first. The White House sniffed contemptuously at the Sputnik and said America had no wish to enter a "basketball game" in orbit with the Russians. Eisenhower himself said the Russian satellite "does not raise my apprehensions one iota."

In private, however, many military and civilian officials were close to panic. All kinds of frightful possibilities came to mind after the first Sputnik went up. If the Russians could put a satellite in orbit, then they might send atomic bombs into space as well, and bring them crashing down on American cities at will. And what if the Soviets should start sending *men* into orbit? The thought of a sky crisscrossed by manned Russian spaceships, carrying reconnaissance cameras and perhaps even nuclear weapons—on the model of von Braun's orbiting missile platform, proposed five years earlier—was enough to make Pentagon generals tremble.

Those apprehensions were confirmed less than a month later. The Russians were indeed planning to send up astronauts. On November 3, 1957, the Soviets launched a second Sputnik, carrying, as in von Braun's plan, a test animal—a dog named Laika—which could only be the surrogate for a spaceman.

Von Braun and his co-workers at Huntsville, who had been experimenting with liquid-fuel boosters capable of putting payloads in orbit as well as carrying H-bombs across the seas, were confident they could launch a satellite at any time, given a few weeks to prepare. In fact, they could have done so years earlier, had anyone in Washington given the approval.

There had been a plan in the mid-1950s to launch a tiny American scientific satellite, equipped with Geiger counters and other instrumentation, to study conditions above the atmosphere. Dubbed MOUSE, for Minimum Orbital Unmanned Satellite, Earth, it was ignored by the federal government. Now something like MOUSE seemed the best bet to become the first U.S. earth satellite; and von Braun, seeing the public outrage over Russia's space triumphs, thought he could get his mandate from Washington. In ninety days, he said, he and his colleagues at Redstone could put an American satellite in orbit, using an Army missile.

A few days after Sputnik II achieved orbit, von Braun received the go-ahead from the Defense Department. The Navy was first to the launch pad, however, with its own rocket, the Vanguard.

A slender vehicle like a giant awl, Vanguard had a disastrous début. On December 6, 1957, two months and two days after Sputnik I stunned the world, Vanguard was launched. Just after lift-off, the first stage malfunctioned, and the rocket collapsed in an enormous fireball. The flight lasted one second. "Kaputnik," one newspaper called Vanguard. "Pfftnik," said another. Now it was the Army's turn.

The Army satellite was named Explorer I. Just under a meter in length and about as wide as a coffee can, Explorer was equipped with a variety of instruments like those which would have gone into MOUSE.

January 29, 1958: five days short of von Braun's promised ninety-day deadline, Explorer I stood ready for launch at Cape Canaveral, Florida. The launch vehicle was a liquid-fuel Juno rocket.

Winds were too strong to permit a launch, and the test was postponed until the following night. Again winds prevented a lift-off. It was January 31st before the weather was right, and on that date, at 10:45 P.M. Eastern Standard Time, the Juno's engine was ignited. Everything went according to plan: all stages went off on schedule; then there was nothing to do but wait and see if tracking stations around the globe would pick up Explorer's signal, telling the world that the Army satellite had attained orbit.

A few seconds past the predicted time, Explorer's electronic voice was still unheard, but half a minute later, the satellite's beeps started coming through loud and clear.

Explorer had reached an orbit marginally higher than expected, which accounted for the brief but heart-stopping delay.

The Soviets no longer had the skies to themselves. In the days of national euphoria that followed Explorer I's success, the concept of a military man-in-space program was very much on the government's mind, as indicated by a report published under the title *The Next Ten Years in Space, 1959–69,* which was the summary of a 1958 study by the Select Committee on Astronautics and Space Exploration. The committee had asked several dozen leading scientists, engineers, military men, and executives of the aerospace industry for their views on what the U.S. and Soviet space programs might accomplish in the coming decade, and what American space policy should be. Their remarks were enlightening.

The report in some parts seems incredibly naïve and impractical today, such as a prediction that mail-carrying rockets could make possible the delivery of letters from Europe to America on the same day that they were written. It is also full of optimistic forecasts of discoveries that might be made by rocket-launched spaceships and orbiting observatories. One rather visionary scientist went so far as to suggest that spaceships might approach the velocity of light by the end of the 1900s.

Most striking about the report, however, is its recurring military theme. Brigadier General H. A. Boushey, Director of Advanced Technology for the Air Force, took it as a matter of course that "military spacecraft will police the vicinity of the earth to prevent the use of space for aggressive purposes"; and scattered all through *The Next Ten Years in Space* are proposals for new space-borne weapons systems.

Eugen Sänger, then Director of the Institute of Jet Propulsion Physics at the Technical University of Stuttgart, contributed a paper titled "The Future of Space Flight," in which he described a possible ASAT and antimissile device that sounded much like the old death rays of science fiction. Sänger imagined a "stationary ultraviolet searchlight" that would fire high-energy beams of radiation which might be capable of destroying flying objects "up to a distance of several hundred miles in a fraction of a second." He expressed a desire that such a formidable weapon would "lead to the ultimate elimination of all warlike tendencies from research in aeronautics and astronautics." (Only a decade and a half after Sänger wrote that paper, the U.S. and the Soviet Union were working on devices much like his sharpshooting searchlight, for knocking out each other's space systems. Whether or not these weapons will chase "warlike tendencies" from aerospace research remains to be seen, but at the moment prospects look less than hopeful.)

House Document No. 115, as the report was officially designated, was, in short, a rough blueprint for war in space. And, as one might expect, Walter Dornberger was prominent among the contributors to the study.

A military career man, Dornberger naturally saw space as a new arena for warfare, and he stated flatly that the goal of the United States should be "the military utilization of space . . . in the fastest possible way." He favored, among other things, a swarm of orbiting atomic bombs that could be guided down to targets in the Soviet Union. Dornberger had proposed such a system as early as 1948, and in *The Next Ten Years in Space* he predicted that "we will see within the next ten years the following space vehicles in our weapon arsenal:

"1. Reconnaissance satellites, optical, radar, infrared and ferret, automatic.

"2. Reconnaissance and bombing space gliders, manned.

"3. Unjammable weather stations, automatic.

"4. Unjammable communications satellites, automatic.

"5. Military maintenance, supply and rescue ships, manned.

"6. Antisatellite space weapon system, manned and automatic.

"7. Satellite defense systems, automatic.

"8. Bombing devices, automatic.

"What these weapons systems will finally look like, up to which altitude they will operate, I don't know yet, but I know we have to have them sooner than our potential enemy."

Some of the systems Dornberger described were still far in the future, but one of them—the reconnaissance and bombing space glider—was already on the drawing boards, thanks largely to his efforts on Project Bomi.

FOUR

"We Expect Threats"

ANTI-GRAVITY spaceships, real flying saucers, moonships as big as ocean liners—these were a few of the schemes for space travel that were hatched in the giddy months just after the launching of the first U.S. and Soviet satellites.

One of the most ludicrous sideshows in the space race during the late 1950s was the much-publicized "flying saucer race" between the United States and Russia. Flying saucers had been a public craze in the United States for the previous decade, ever since a pilot had reported seeing luminous discs in the sky near Mount Rainier in Washington State. Saucer mania was further stimulated by the release in 1951 of Robert Wise's film *The Day the Earth Stood Still*, which opened with a brilliantly photographed sequence showing a disc-shaped alien spaceship landing on the Mall in Washington, D.C. Soon everyone was seeing saucers in the skies, and police departments all over the United States and Canada learned to brace themselves for an annual silly season of saucer sightings. The season tended to last from late spring to early autumn, when large numbers of persons were outdoors looking at the sky for various reasons.*

So it should have come as no surprise that the Russians —who, in the 1950s, had claimed to have invented nearly everything else—dropped hints around that time that they had developed an anti-gravity flying saucer, too. In the prevailing atmosphere it seemed as if Americans would believe practically any fantastic story; perhaps they would swallow this one, too.

Anti-gravity engines have fascinated Russian rocket engineers ever since the days of Konstantin Tsiolkovsky, a nineteenth-century schoolteacher who was among the first to work out the theory of spaceflight. When he was in his teens, Tsiolkovsky imagined one evening that he had found a means of neutralizing gravity, and he wandered

---

* There had been another wave of UFO mania in the United States just before the turn of the century. Rather than saucers, people were seeing airships like the early dirigibles being developed in Europe. According to one report, a mysterious airship stopped at a man's home one evening, and the vessel's occupants demanded a pot of coffee and a plate of egg sandwiches!

through the streets of Moscow all that night in a daze, staggered by the implications of what he thought he had discovered. In the cold light of dawn, however, he saw he had been pursuing a chimera, and gave up anti-gravity in favor of conventional chemical-fueled rockets.

But Tsiolkovsky's dream was not forgotten, and about the time that Sputnik I went up, Dr. Kirill P. Stanyukovich, a math teacher at the Moscow Higher Technical School, said he expected "forthcoming advances" in Soviet research on ways to overcome gravity. Stanyukovich was an internationally known and respected mathematician and physicist, and his words carried considerable weight in the West. He based his speculations on his own theory of gravitational energy, which, he said, took the form of radiation called "gravitational quanta," or "gravitons." Stanyukovich speculated that this radiation would be stronger at higher temperatures and weaker at lower temperatures. So, in theory, one could reduce the gravitation of an object by chilling it. And at absolute zero, the temperature where all atomic and molecular motion ceases, there should be no gravitational potential either. According to Stanyukovich's ideas, if one tried to walk on a planet that had been chilled to absolute zero, one would fly off into space upon trying to take a step forward—because there would be no gravity to counteract the thrust from one's leg muscles.

Stanyukovich suggested it might soon be possible to manufacture "weightless matter" and build aircraft and spaceships out of it. He called such vehicles "graviplanes."

In 1957, U.S. scientists and military people were so spooked by recent Russian achievements in space that *any* claim of this kind, no matter how preposterous it seemed, was scrutinized to see if perhaps there was some truth in it. In 1961, *Missiles and Rockets* magazine ran two articles on the prospects for anti-gravity spaceship drives. The articles concluded that while it was theoretically possible to cancel out gravity and make a vehicle weightless, a practical anti-gravity engine was a long way from realization.

Just how far from realization it was, the Soviets themselves revealed when they released photos of what they claimed was a graviplane in flight. At first glance, it looked much like the classic model of a flying saucer. It was disc-shaped and had a clear canopy over the cockpit, and it appeared to be floating above the ground with no source of motive power. On close examination, however, the photos turned out to be a hoax. What gave it away was the tall vertical stabilizer at the tail end of the craft. A true anti-gravity aircraft would not need such control surfaces, because they are required for maneuvering only if the vehicle has to drive itself forward through the air to generate lift; since a true graviplane would have no need for aerodynamic lift there would be no use for a vertical stabilizer, either.

Not to be outdone, the U.S. Army announced that it was also working on a flying saucer. Like the Russian graviplane, the Army saucer was an anticlimax when finally unveiled. Instead of a sleek disc whizzing through outer space, the Army saucer was nothing more than a big ducted fan about the size of an automobile. Intended for use as a scout vehicle, it never rose more than a foot or two above the ground, and was finally retired to the Army Transportation Museum at Fort Eustis, Virginia.

Almost as fantastic as the flying saucers were the other

spaceship designs conceived in the late 1950s. One was Antares, a nuclear-powered space rocket that was capable, on paper, of landing some two million kilograms of cargo on the moon—enough to found a lunar base, which could then be supplied with more material on future flights. Militarizing the moon was a popular notion around this time, even though armed forces on the moon would have served little or no useful purpose. The moon was too distant to make it practical as a reconnaissance outpost; satellites in low orbit could do that job much more quickly and cheaply. Nor was there any sound cause to plant missiles on the moon for attacks on the Communist world; a missile launched from the moon would have to travel 350,000 kilometers to hit anything on earth, whereas an earth-based ICBM would have to fly only a small fraction of that distance.

Still, moon fever was powerful in the early days of the U.S.-Soviet space competition, and gave rise to a spaceship design even more awe-inspiring than Antares, called Aldebaran. A spacegoing behemoth bigger than the ocean liner S.S. *United States*, Aldebaran looked somewhat like a rocket-powered moth. It had a cylindrical fuselage the size of a battleship, a hemispherical nose, and a huge bell-shaped exhaust nozzle at the stern. On either side of the fuselage were the "wings"—actually great curved floats designed to support the mammoth spaceship on the water. Aldebaran was meant to take off and land at sea, there being no facilities on land that were big enough to handle it. Like Antares, Aldebaran was planned for the earth-moon run; on every trip, the spaceship could haul enough equipment and personnel to colonize a fair-sized planetoid. And—who could tell?—perhaps the asteroid belt would be opened to settlement someday. In that event, there would be need for a superspacecraft like Aldebaran.

Propelling ships that size would be a challenge, to put it mildly. Conventional chemical fuels—kerosene, liquid oxygen, and the like—were out of the question, because they did not pack quite the punch needed to lift thousands of tons of payload into space on a single launch. The answer? Nuclear energy.

One plan was simply to detonate small atomic bombs just astern of the giant spacecraft, and shove it into orbit in the manner of Orion. Another possibility was to touch off a bomb within a spherical chamber astern of the spaceship; the blast would escape through a hole in the sphere and provide enough thrust for takeoff. A more sophisticated scheme involved using heat from a nuclear fission reactor to vaporize liquid hydrogen carried in large insulated tanks on board the spacecraft. The supercold liquid would expand violently upon heating, and would drive the rocket up into space. This method of propulsion had one big advantage over exploding A-bombs: no hazardous fission products, such as radioactive strontium, would be released into the atmosphere.

Fantasizing about nuclear rockets and lunar cargo runs was fun, but any plans were destined to remain in the blueprint stage. The U.S. had its hands full, in the late 1950s and early 1960s, merely building liquid-fuel rockets capable of orbiting anything bigger than a sack of potatoes. Before the U.S. could send up spacemen, it would have to develop a booster capable of lifting a multi-ton spaceship out of the atmosphere. Therefore America concentrated on the Titan and Atlas rockets, ICBMs designed to double as spaceship launch vehicles.

Meanwhile, the "specter of the spaceman," as the *New Republic* called it, was becoming a hot political issue in the U.S.—particularly in the 1960 Presidential campaign. The Republican candidate, Vice-President Richard Nixon, was quiet at first about the sensitive subject of militarizing space. The GOP line, held over from the Eisenhower years, was that there seemed no urgent need for a space-arms escalation. Nixon evidently hoped that if he ignored the issue, it would go away.

The issue was hard to ignore, however, after Democratic candidate John F. Kennedy accepted an invitation from *Missiles and Rockets* to spell out his views on the armed forces' role in outer space. In plain terms, Kennedy affirmed his commitment to keeping the United States ahead of the Soviets in what he called the "vital race" for supremacy in orbit. Though he refused to tie himself to specific dates for accomplishing such projects as manned space stations and nuclear space rockets (the timetable for them, he thought, should be "elastic"), Kennedy proclaimed an expanded space effort to be part of his New Frontier and promised he would not allow the Kremlin to achieve pre-eminence on that front.

It was a stirring pledge, and it forced Richard Nixon to make his position known. Shortly after Kennedy's article appeared in *Missiles and Rockets*, the magazine printed one by Nixon in which he defended the Eisenhower administration's record on space policy as "creditable," and pointed out ways in which the United States was actually ahead of Russia. The U.S. had launched more satellites than the Soviets had. Ours had stayed up longer than had theirs. The U.S. seemed to be well ahead in recovery of satellite payloads. In short, Nixon said, America was not losing the strategic space race with Russia, and indeed had overcome their formidable head start. But while Nixon admitted the military had legitimate business in space, he also stood up for the Eisenhower administration's controversial doctrine of "freedom of space"—the astronautic equivalent of "freedom of the seas"—which held that space should be an international "ocean," and not the domain of the earthly powers with the biggest weapons. That doctrine had come under heavy fire from the U.S. military and the aerospace industry; it struck them as an open invitation for Moscow to push for control of outer space. Now Nixon's defense of freedom of space put him at a political disadvantage, for Kennedy, capitalizing shrewdly on the beat-the-Reds spirit popular in the land, came out of the debate looking like the candidate who would do more to halt the perceived Red peril in the skies.

Kennedy's views were backed by the brass in Washington. Air Force Lieutenant General Roscoe C. Wilson, in a 1960 interview, made the ominous prediction that "we have no time at all" to lose in the military space race with the Russians, and added that he felt the U.S. must speed up its preparations for war in space "exponentially." Air Force General Curtis E. LeMay, in testimony before the House Appropriations Committee some months after Kennedy's article was published, said: "We expect threats to emerge. Therefore, we should be doing some work in this area [the militarization of space] so that we will be capable of countering the threats if they appear or anticipate threats that may appear."

Then, in 1961, a threat from Communist spacemen did indeed seem imminent when the Soviets succeeded in putting their first man in orbit. Yuri Gagarin was launched aboard his Vostok I spaceship from the Tyuratam (also known as Baikonur) space complex in Soviet Central Asia on the morning of April 12th. He spent approximately ninety minutes in orbit, completed one revolution of the earth, and returned safely in his re-entry capsule, landing in Kazakhstan, in the southwestern U.S.S.R., at 10:55 A.M.

Once again, the Russians had demonstrated their superiority in space. As the first man to orbit the earth in a space vehicle, Gagarin became the No. 1 asset for the Soviet propaganda machine. His boyish good looks were reproduced on magazine covers all over the world, and he was touted as an inspiration to Socialist youth. Nixon's defense of the U.S. record in space suddenly sounded hollow, and the signs of Russian supremacy in orbit again began to make Americans edgy about their nation's security.

The U.S. reaction was swift. Less than two months later, after Kennedy had been elected to the Presidency, his administration took the first bold steps toward establishing a manned military space force. Gone was Eisenhower's lack of "apprehensions."

The Kennedy administration stood for expanding the Pentagon's domain into space, and asked Congress for money to fund several projects whose military potential was plain:

Giant liquid- and solid-fuel rocket boosters, perfect for putting large payloads such as manned military space vehicles in orbit.

The Titan II liquid-fuel rocket, an advanced version of an already developed ICBM, for Air Force space endeavors.

Apollo, America's man-on-the-moon project, to which Kennedy committed the U.S. shortly after his inauguration. Though officially a civilian project to be carried out through NASA, Apollo was widely viewed as a test program for hardware that could be applied to DOD's space jobs.

Nonetheless, Kennedy took care in his public statements to avoid any direct reference to U.S. military activities in space. In a speech to Congress on May 25, 1961, for example, he never mentioned the militarization of space when he asked the legislators to vote more than $600 million in appropriations for the space program. Neither did he touch on the fact that more than $75 million of that sum was earmarked for the Pentagon. The closest he came to doing so was when he said America's "leading role in space achievement" might determine "our future on earth."

By softening the administration's war-in-space rhetoric, Kennedy seems to have been trying to invite Russia's co-operation in limiting the arms race in orbit. He made it clear in a 1962 speech delivered in Houston that the U.S. did not wish to see space "filled with weapons of mass destruction, but with instruments of knowledge and understanding." He went on to say:

"I do not say that we should or will go unprotected against hostile misuse of space, any more than we go unprotected against the hostile use of land or sea. But I do say that space can be explored and mastered without feeding the fires of war, without repeating the mistakes that man has made in extending his writ around this globe of ours."

At a press conference in August of that year, the President summarized the administration's views this way:

"There is a great interrelationship between the military and the peaceful use of space. But we're concentrating on the peaceful use of space, which will also help to protect our security if that becomes essential." In other words, the U.S. intended to pursue space projects, in the name of peace, that might just as conveniently be used in war if necessary.

Not everyone in Washington tiptoed around the issue of arming outer space. Senator Stuart Symington suggested the United States would be committing "unilateral disarmament" if it did not put nuclear bombs in orbit, in accordance with Dornberger's plan. Symington, a former Air Force Secretary, complained that America was "denying the military prompt development of the best possible . . . weapon in this new militarily so important arena." Air Force General Bernard A. Schriever, whose fervent backing for military space projects earned him the unofficial title "General of Outer Space," told an audience at Mississippi State University on April 10, 1962:

> "If the Soviets should attain a significant breakthrough in space technology, they could achieve two decisive military capabilities. They could develop spaceborne weapons systems that could attack targets in the Free World. And they could deny other nations access to space—even for purposes of scientific research. Soviet achievement of either of these capabilities would pose a grave threat to our national security."

To meet the Soviet challenge in orbit, the U.S. Air Force was working on a variety of space weapons. One was Saint, short for SAtellite INTerceptor, an early ASAT system.

The Air Force made good use of psychology in naming

these weapons. Who, for example, would dare commit the sacrilege of opposing a Saint? There was also something code-named Early Spring, because of an old Washington proverb that there are two things no one can rightfully be against: motherhood and an early spring. Even Walt Disney characters were represented in our space-weapons roster. One anti-missile project, which would have used 800 to 3,600 satellites armed with interceptor missiles to shoot down Russian ICBMs, was dubbed Bambi—Ballistic Missile Booster Interceptor. Russian military space systems, by contrast, were described officially as "satellite killers" and the like.

Part of the fun of designing a new weapons system, it seems, was inventing a clever name for it. One of the memorable ones was Dyna-Soar, a contraction of Dynamic Soaring. Dyna-Soar was a rocket-propelled space glider, an outgrowth of the Dornberger Project and an heir of the Sänger-Bredt bomber of World War II. It was intended to serve as the reconnaissance and bombing vehicle Dornberger described in House Document No. 115.

Every spacecraft has a personality. Some look clean, efficient, and utilitiarian, as the Apollo moonships did. Some look comical; seen from the front, the Gemini two-man space capsule resembled a mournful Cyrano. Dyna-Soar looked downright sinister.

Known officially as the X-20, Dyna-Soar was black as space itself, except for its insignia. About 10 meters long and 6.5 meters in wingspan, it resembled a huge obsidian arrowhead. The narrow delta wing, raked back at an angle of almost 80 degrees, turned up sharply at the tips to form two vertical stabilizers. The wing was set low on the body so as to give the spaceship a flat belly for maximum lift on re-entry—a feature of the German antipodal space bomber. The boxlike fuselage, 2.5 meters high, tapered into a shovel nose in front and flared outward at the tail end to accommodate the nozzles of the X-20's rocket engines. There were two little rudders on the trailing edges of the vertical stabilizers, and two elevons (horizontal control surfaces) on the delta wing.

At first glance, Dyna-Soar appeared to be blind because no cockpit windows could be seen from the outside. They were covered with a protective metal shield which was to remain in place until after re-entry, to keep the windows from cracking and shattering under the terrific heat and pressure of re-entry. During lift-off there was no need to see outside, and in orbit the pilot could use radar for tracking and rendezvous.

Dyna-Soar's re-entry would differ slightly from that planned for the antipodal bomber. The Sänger-Bredt spaceship would have made several successive dips into the atmosphere before re-entry, to slow the spaceship for its final descent without heating the hull too severely in the process. Dyna-Soar, however, was made of tougher stuff. Thanks to a special ceramic heat-resistant nose cap, the craft could survive a single long glide down from orbit without melting its skin. (This nose cap was an amazing achievement. It radiated away far more heat than it absorbed, had a melting point well above 2,000 degrees centigrade, and could withstand the direct blast of a engine for periods longer than a descent from orbit would take.)

Like the Sänger-Bredt bomber and the A-9, the X-20 was intended to return under a pilot's control, and land as a high-speed aircraft would. It would be lifted into space by a liquid-fuel Titan booster; then, after its business in space was completed, the pilot would bring the craft back in the same re-entry attitude as the German space bombers—nose up and belly forward—while remaining safe and snug in his insulated cabin. In the lower air, he would use the control surfaces to bring the spacecraft down to a three-point landing. Instead of wheels, the X-20 was equipped with wire brushes or, in some models, skis. This kind of landing gear could withstand the heat and pressure of a high-speed landing much better than could inflated tires.

Dyna-Soar was designed to land almost anywhere that the pilot could find a suitably long and flat stretch of pavement. Even a highway would do, in an emergency.

Officially, the X-20's job was research. It was described in DOD handouts as an experimental rocket plane meant to test new materials and building techniques for manned spaceships. Now and then, however, the military talked publicly about its plans for the X-20.

An Air Force undersecretary made the X-20's military mission plain in a speech he delivered at the unveiling ceremony for a scale model of the spaceship in Las Vegas in 1962. "The X-20," he said, "is designed to explode the explored narrow band of speed and altitude into a complete corridor to space—a corridor within which man will be able to exit and re-enter from space. . . ."

It was not long, however, before he brought up the reason for DOD's interest in the X-20. "The X-20 does not represent a vehicle for a specific military job," he said, "but neither did the Wright brothers' original airplane." The implication was clear: this might be only the first of a long line of U.S. military spaceships—in effect, an American Space Force.

The X-20, he predicted, "will be a milestone in the conquest and utilization of space, and the men and women who fly it will be the artisans who will convert space flight from an experiment to a new medium for exploitation—hopefully for peace but, if need be, for the preservation of our national security."

More to the point was an article about the X-20 in a 1963 issue of *Aviation Week,* in which it was said that the spaceship was expected to provide the groundwork for the Air Force to handle any military challenge in space in the years to come. The article listed a number of possible military jobs for the X-20. It might be used to transport men and equipment, or to carry out reconnaissance operations. "Interdiction" was another task on the list—and that word implied that the X-20 might be called upon to fight battles in orbit, or at least to knock unmanned enemy spaceships out of the sky. The article went on to quote a deputy director of the X-20 program as calling the spaceship an "interceptor." Suddenly the X-20 pilots were starting to look less like artisans and more like combat fliers.

Dyna-Soar was only one of several comparable projects the Defense Department was studying in the early 1960s. Another was a space bomber called BOSS/WEDGE, for Bomb Orbital Strategic System/Weapons Development Glide Entry. BOSS/WEDGE was supposed to investigate a variety of missions that manned spaceships might perform in wartime, from bombing to reconnaissance. There was also an advanced Dyna-Soar design called Dyna-Mows. Like BOSS/WEDGE, it would have carried bombs to drop on targets on earth.

America's military man-in-space plans were given

added impetus by Russia's second Vostok flight, on August 6, 1961. Early that morning, while Washington slept, the five-ton Vostok II took off from Tyuratam with Major Gherman Titov on board. Tass, the Soviet news agency, described the launch in purple prose: "Gathering force, it streaks more and more quickly upward like a roaring fiery globe. . . ." Titov remained in orbit for more than twenty-five hours, compared to only an hour and a half for his predecessor, Gagarin, and traveled approximately 609,000 kilometers—more than seventeen times Gagarin's total. Titov became, on this flight, the first person to sleep in outer space. At 6:15 P.M. Moscow time, just short of nine hours after launch, Titov said, "Dear Muscovites, I am going to sleep; you can do as you like." At two o'clock the next morning, he awakened. Eight and a half hours later, Vostok II was back on the ground. Titov and his spaceship touched down near Engels, a factory town on the Volga River about 630 kilometers southeast of Moscow. Titov was in good condition and reported that all his ship's systems functioned perfectly. An exuberant Premier Nikita Khrushchev placed a phone call to Titov and burbled, "I kiss you and embrace you by telephone for the time being." Later, at ceremonies honoring Titov in Moscow, Khrushchev had the opportunity to kiss and embrace him in person.

There were no celebrations in America, for Titov's flight only underscored how far ahead the Soviets were in manned spaceflight. Khrushchev made that point time and again in his maddening taunts to the U.S. "The Americans do not launch," he said once, in a scornful reference to the first Mercury suborbital flight in 1961. "Their men only jump upward and fall into the ocean."

To the Air Force, one of the most disturbing things

about Vostok II was Titov's assertion that he had "guided" his spaceship down to a landing. At a Moscow press conference after the mission, Titov said, "My spaceship . . . was very easy to guide." He added, the following day, that a vessel like Vostok II could be directed by its pilot to any spot on the globe. In a sense, he was telling the truth. If launched into a polar orbit, which would carry it over the whole surface of the globe as the earth turned on its axis, then a Vostok could be brought down anywhere on earth. The re-entry would simply have to be timed precisely. Vostok II was not, however, the kind of winged vehicle that could be piloted like an airplane, as the Air Force had in mind for Dyna-Soar. Rather, the Vostok cosmonaut would have to return in a ballistic capsule, much like the U.S. Mercury spaceship but bigger.

Still, Titov's fleeting reference to "guided" spacecraft was enough to make Washington extremely uneasy. Almost nothing was known for certain in the United States about Russia's manned space vehicles (at one point U.S. analysts were reduced to studying Soviet commemorative postage stamps, hoping the fanciful spaceships depicted on them would yield some useful intelligence); and if the Russians had developed a space vehicle that could maneuver on its way down, then quite possibly they had their own Dyna-Soar already!

That was the gist of an article in a 1960 issue of *Missiles and Rockets* titled "Russian Dyna-Soar Flying?" The article reported rumors that the Soviets were conducting flight tests on a piloted "semi-ballistic missile" called the T-4A. According to information reaching the West, the T-4A was "quite similar in design" to Eugen Sänger's antipodal space bomber.

In fact, the Soviets had taken an interest in Sänger's

design shortly after the German surrender in 1945. Though Sänger himself wound up in U.S.-British custody, a copy of his report on the antipodal bomber, "Über einen Raketenantriebe Fernbomber" ("Concerning a Rocket-Propelled Long-Distance Bomber"), fell into Russian hands. A translation of the document eventually landed on the desk of Stalin himself, and was said to have interested him greatly. Having just seen his armies miss out on the treasure trove of hardware and information at Peenemünde, he looked upon the Sänger design as a way to make up for that loss.

So in April, 1947, the head of the Russian military missile effort, Colonel Grigori A. Tokadi, was summoned to the formidable presence of Stalin and the Politburo. At Stalin's order, Tokadi spoke in detail on the subject of the antipodal bomber, painting a generally pessimistic picture of it. The Soviets, Tokadi pointed out, knew too little about materials and design to enable them to build such a complex weapon successfully.

That remark seemed to annoy Stalin. "We will have to learn from the Germans, is that it?" he reportedly asked Tokadi. Though reluctant to answer, Tokadi was pressed by Stalin and finally told him that the Germans seemed to have given more thought to the military uses of rockets than the Russians had.

At last the conference was over. Tokadi survived Stalin's displeasure and was ordered to find Eugen Sänger and bring him to the Soviet Union—of his own free will if possible, but by force if necessary. Tokadi was accompanied by General I. A. Serov (the same officer who had incurred Stalin's rage a few years earlier by letting the Peenemünde team slip through his grasp) and by Stalin's son Vasili, who, like his father, was a Politburo member.

The three Russians spent several months combing Western Europe for Sänger, but they never found him, which is curious because Sänger was living openly in Paris at the time. They also searched for Sänger's co-worker, Dr. Irene Bredt, but were equally unsuccessful at finding her.*

After Tokadi, Serov, and Vasili Stalin returned empty-handed from their mission to locate Sänger, the Russians, it was learned later, reverted to bureaucratic form and set up a committee to study the space-bomber proposal further. There the plan was put in limbo, only to be resurrected more than two decades afterward.

It is unlikely, then, that a Russian Dyna-Soar was undergoing tests in 1961. The T-4A sounded so much like the Sänger-Bredt bomber—it was even supposed to be launched down a long steel track by a rocket-propelled catapult, as in the German design—that the Russians may have just taken the original German report off the shelf, changed the specifications slightly, and leaked the rumor of a Soviet "antipodal semi-ballistic missile" to the West, in hopes of frightening the United States still further. In any event, there was no solid evidence until the 1970s that the U.S.S.R. was working on anything like the Dyna-Soar or the Sänger-Bredt bomber.

The U.S. could probably have launched its Dyna-Soar in the early 1960s. The technology was available, or well on the way to being developed. Initial drop tests from B-52 bombers were scheduled to begin in 1963, with the

---

* She married Sänger after helping him design the antipodal bomber, and changed her name to Sänger-Bredt.

first unmanned flight the following year and a manned orbital mission in 1965. Dyna-Soar, however, was doomed to extinction before it left the launch pad. In December, 1963, the project was canceled.

The X-20 was a victim of simple haste. In the frantic rush of the U.S. to catch up with Soviet achievements in space, the X-20 wound up competing with the semi-military Gemini capsule, which had many advantages. Gemini was nearer completion; it was much more versatile, and could be used on a wider variety of missions; and it could do virtually any jobs the X-20 could, without extensive modification. There had been design problems with the X-20. Its weight had increased. It could carry only a small payload. In short, the X-20, as first conceived, was obsolete by the time it was ready for flight.

A few scaled-down versions of the X-20, similar in materials and design but unmanned, were launched in a series of tests from 1963 to 1965, with mixed results. These mini-spaceships were Dyna-Soar's last gasp in the 1960s. It would be another decade before the Dyna-Soar concept would rise, phoenixlike, from oblivion, in a different and much more subtle form.

FIVE

"Let Our Foes Know"

BETWEEN Scylla and Charybdis—that is where the issue of America's military space program had put the Kennedy administration by mid-1962.

Soon after setting the program in motion, Kennedy found himself under pressure from the armed services—especially the Air Force, which had the most to gain—to accelerate it still further. The Air Force position was stated succinctly by one general, who said that in the next few decades the decisive battles "may not be sea battles or air battles but space battles."

As if to underscore the Air Force message, the Soviet Union in 1962 staged yet another space spectacular. This time the Russians sent up not merely one manned spaceship, but *two*. On August 11th, Andrian Nikolayev was launched into orbit in his Vostok III spaceship. The following day, he was joined by Pavel Popovich in Vostok IV.

The highlight of the two-ship mission was a rendezvous in space. Vostok III and Vostok IV passed within five kilometers of each other—almost shouting distance, had there been any air to shout in.

It was clear that Russia had nearly perfected the techniques required for docking in space. Once the Soviets had that maneuver down pat, they would have overcome one of the biggest obstacles to putting a manned military space station in orbit.

Nikolayev and Popovich both returned to earth on August 15. Shortly afterward, Defense Marshal Rodion Malinovsky used the occasion to boast: "Let our foes know what technology and what militancy are in the possession of Soviet power!"

When Malinovsky spoke, the West listened; as Soviet Defense Minister, his was the voice of the U.S.S.R. armed forces. About the size and shape of a grizzly bear, he had thick black eyebrows and bushy white hair, and a fearsome scowl. When Malinovsky rumbled about Russia's "militancy" in space almost in the same breath as he mentioned the August space mission, shocks were felt throughout the Pentagon.

With every Soviet space mission, the Air Force saw its arguments for a U.S. spacewar program strengthened. Time was wasting, the White House was warned, in view of the long lead time required for mobilizing any new weapons systems. As weapons grew ever more complex, delays of a decade or longer between commission and completion of a project were becoming commonplace. Missiles and spacecraft ordered in 1962 might not be ready for delivery until 1967 or even the early 1970s. *Now,* said the Air Force, was the time for America to launch a crash military space program. Tomorrow might be too late.

Kennedy had no wish to be portrayed as "soft on Communism," either on earth or in the skies. Yet, because it seemed imperative that the U.S.-Soviet arms race be brought under control, he could not afford to take a gung-ho attitude toward military space projects. The arms race was expensive and dangerous for all parties concerned, as demonstrated by the Cuban missile crisis of October, 1962.* Belligerent rhetoric, the administration knew, was only likely to make the situation worse.

Caught between the demands of the armed forces and the pressing need to cool East-West tensions, Kennedy took the middle road. He proceeded with Pentagon space projects slowly, hoping to avoid provoking the Soviets into a new arms spiral in orbit.

But a government program—especially a military one —is difficult to stop once it is set in motion. And the more money that was allocated for DOD's space effort, the more momentum the Pentagon's plans seemed to gather.

Dyna-Soar had merely whetted the Air Force appetite for manned military spacecraft. Soon after Kennedy's State of the Union Message in May, 1961, in which he

* Discussed in Chapter 6.

committed the United States to a lunar landing "before this decade is out," and NASA was planning manned orbital missions as "stepping stones" to the moon, the Defense Department began to muscle in on Project Gemini.

Gemini, named after the constellation of the twins, was a two-man spacecraft and the successor to the tiny Mercury space capsule. Launched into orbit by a Titan booster, Gemini was to be a flying test bed for hardware that would go into the Apollo moonships. Though officially Gemini was a non-military project run by the civilians at NASA, DOD saw Gemini as a golden opportunity.

Ever since von Braun suggested putting a military platform in orbit, the armed forces had wanted one for purposes of espionage. However, the principal obstacle to building a DOD space station was ferrying men and material back and forth to it on a regular basis.

Dyna-Soar had been intended for that job (among others), but enthusiasm for the spacecraft was beginning to wane in Washington by 1962. Now the Air Force had its eye on Gemini. As long as NASA was going to send up this two-man spaceship, the Air Force reasoned, why not give it some military duties too?

Gemini was a different proposition from the Dyna-Soar, which was a big unknown; no craft like it had ever been sent into space and back before. Gemini, on the other hand, was a more familiar design. It was a ballistic capsule whose behavior on re-entry could be predicted easily. It would require few modifications to serve a military role in space, whereas Dyna-Soar would need extensive alterations if it was to serve as a supply vessel for space stations as well as an interceptor and research vehicle.

The Defense Department kept Gemini astronauts busy

on their early flights. One of their jobs was to determine how effective manned satellites might be for reconnaissance. Gemini 4, launched on June 3, 1965, carried up a hand-held camera. The resulting pictures were so revealing that DOD sent up a pair of telephoto lenses on Gemini 5, which was launched on August 21st. The photos taken with these lenses showed so much detail that only two of them were ever released, and those pictures were deliberately blurred by DOD for security reasons. Even the pictures taken from Gemini 5 with an ordinary camera—which were released unchanged—revealed a startling amount of detail. Highways showed up boldly without magnification, as did buildings. In theory, they should have been invisible from the Gemini's 140-kilometer altitude; but there they were.

As the Gemini astronauts clicked away in orbit, and the Russians accused the U.S. of using its Gemini for spy missions against the U.S.S.R., NASA was studying space-station designs. The space agency planned to start small, with a four-man platform. The next step would be a much larger facility called MOSS, for Manned Orbiting Space Station. Several designs were considered for the platform. One was the familiar torus, first imagined by von Braun: an "inner tube" spun slowly to create a semblance of gravitation by centrifugal force. Occupants of the station would walk on the inside of the tube's outer rim, feet pointed outward from the station's hub.

A slightly different plan was the "radial" design. It called for a central hub to have several cylindrical living and working modules pointing outward from it, like the arms of a starfish. The hub, where centrifugal force was least, would be a zero-gravity area and the site of docking ports, because a spaceship could approach most easily where the station's motion was least.

NASA's Langley Research Center in Hampton, Virginia, came up with a design much like that of the later Skylab: a cylindrical station with large winglike solar panels on either side, and several docking ports for Gemini spacecraft, as well as for unmanned robot supply vehicles. To save money, part of the station would be built out of an upper stage of a Saturn rocket.

What the Air Force had in mind was the Manned Orbiting Laboratory (MOL). Similar to the NASA-Langley plan, MOL would be a manned military space station with a variety of missions, including orbital reconnaissance. MOL could also serve as a military communications satellite, linking Washington directly with commanders in the field. Another projected task for MOL was mapping; cameras on the orbiting laboratory could map vast areas of the earth's surface with minimal distortion.

Soon NASA and the Air Force had reached an agreement. The Air Force would let NASA put experimental packages on some of its spacecraft, and NASA would help the Air Force with the development of manned military space stations.

NASA's biggest contribution to the Air Force's man-in-space plans would be to aid in developing a military version of the Gemini space capsule, variously known as Gemini X, Gemini B, and Blue Gemini—Air Force blue, of course.

Blue Gemini was almost identical to the civilian Gemini, except that in the latter the hatch was in the *side* of the spacecraft, whereas in Blue Gemini a hatch was cut in the heat shield. The shield would be facing the MOL when the two craft joined together, so that the astronauts could crawl directly from the Gemini to the lab.

Much more than this the public simply does not know, because many of MOL's specifications remain government secrets. Secretary of Defense Robert McNamara was extremely tight-lipped about the details of MOL when he officially revealed its existence at a Pentagon news conference in 1963.

McNamara began by saying, "My purpose today is to announce the cancellation of the Dyna-Soar project." Dyna-Soar, he told the press corps, had suffered from a "very narrow" objective, and that was one reason it was discontinued.

McNamara went on:

> "We have broadened the objective; the objective of the Gemini X/Manned Orbiting Laboratory program will be to explore operations in space using equipment and personnel that may have some military purpose."

One reporter asked, "What is the military application of the MOL?"

"The military purposes include potential navigation aids, meteorological projects, and other classified projects," McNamara replied.

In answer to a question about whether the MOL would make the Russians nervous, McNamara said that both the United States and the U.S.S.R. had "supported the United Nations resolution stating it is not our purpose to utilize space as an environment in which to develop the capability for use of large-scale mass destruction weapons."

In that United Nations resolution (adapted unanimously on December 20, 1961) concerning the military uses of outer space, the United States and the U.S.S.R. had announced their desire to keep space free of nuclear bombs and other such weapons; neither side had ever promised, however, to ban military systems *per se* from orbit. This loophole, later codified in the International Outer Space Treaty of 1967, left the way wide open for the launching of all manner of spaceborne weapons systems, as long as they contained no nuclear explosives.

NASA, the nominally peaceful U.S. space agency, seemed to welcome DOD's partnership in exploiting outer space, and freely admitted that DOD would be using the MOSS for military duties. NASA's Manned Spacecraft Center in Houston said in the spring of 1963 that NASA had in mind a space station designed for, among other jobs, "early-warning reconnaissance, surveillance, and defensive and offensive possibilities."

And what did NASA mean by "offensive possibilities"? The space agency was rather close-mouthed on that subject, but it was becoming clear that the Air Force was making inroads on the civilian space effort—with NASA's willing cooperation.

One thing helped the Air Force greatly. It had no effective opposition in the administration. As stated in a 1962 article in *Science* magazine, aerospace research and development was distributed among so many different firms and branches of the government that "there is no one man or small group of men who can command the public's attention on the policy questions of space." There were no Robert Oppenheimers, no Leo Szilards in the space program to advance the question: should we be doing what we're doing?

So the Air Force met little resistance to its campaign

for military space projects. Indeed, it met virtually no resistance at all, except from McNamara's budget shears (he was a ruthless cost-cutter—a legacy of his years as an auto industry executive) and from White House reluctance, for political reasons, to push the spacewar program too strongly.

Moreover, the Air Force had influential friends on Capitol Hill. Among them was Senator Howard Cannon, who implored the Kennedy administration to give the military free rein in space.

As a member of the Senate Aeronautical and Space Sciences Committee, Cannon had a hand in determining American space policy; he urged the government to "remove the inhibitions under which the Department of Defense is now laboring and allow [the military to] develop both near-term space systems and the technology for the future."

Thus, by the middle of 1963, the Air Force and its many friends in Washington had brought national space policy about almost 180 degrees. During the Eisenhower years, the government claimed there was no place for the military in orbit. Now, less than two and a half years after Eisenhower had retired to his Gettysburg farm, Kennedy's Secretary of Defense was asking the Air Force to investigate ways of setting up "a method of inspection in space, identification of other satellites in space, rendezvous in space and the capability to neutralize hostile satellites." *Missiles and Rockets* reported that there was a "receptive climate for military space" in the capital, and that "the handwriting on the wall" had given new impetus to the American drive to militarize outer space.

By August, Congress was thinking of formally giving

NASA a military branch. The House Science and Astronautics Committee proposed creating a NASA division with the "responsibility for uncovering and developing military applications from civil space techniques as they evolve." A similar measure had actually passed the House in 1960, but the Senate failed to act on it.

One of the reasons for this legislation was to avoid needless duplication of work between NASA and the Air Force. As it happened, NASA did not get its military scion in a formal sense; but DOD continued to be a powerful presence in the civilian space agency's offices, and in a few years the Pentagon would be using this very same reason —duplication of projects—as an excuse to swallow NASA entirely, leaving the American space program under complete military control.

It seemed only a matter of time before there would be a showdown between the Air Force and the White House over whether to proceed with a vigorous military space effort, or to continue with a policy of comparative restraint and hope the Russians could be persuaded to do the same.

What President Kennedy would have done in that event, no one can say, for on November 22, 1963, he was assassinated.

While the nation mourned the death of its young President, the aerospace industry and all other proponents of a vigorous space program looked upon Lyndon Johnson's succession as an occasion for joy.

During his years in the Senate, Johnson had consistently backed space projects. As Vice-President, Johnson had headed the National Aeronautics and Space Council, which had taken a hawkish view of the U.S. role in outer space. Now, as President and Commander-in-Chief, he had an opportunity to expand the American military presence in space dramatically.

One of Johnson's first decisions regarding space was to proceed with MOL. The initial flight was slated for 1968.

Critics of MOL feared it would merely encourage the Air Force to put bigger and bigger platforms in space, and would engross the U.S.-Soviet arms race still further, but MOL did nothing of the kind—because it never flew. By 1969, the American adventure in Southeast Asia was starting to strain even the huge resources of DOD. McNamara's successor, Defense Secretary Melvin Laird, was being forced to cut MOL expenditures in order to fuel the forces in Vietnam. So, on June 10, 1969, over the bitter protests of the Air Force, whose top officials said the need for a military space laboratory was urgent, the MOL project was terminated.

In fact, there was no need for MOL anymore, at least not in its role as a manned reconnaissance platform. By the late 1960s another military space project had eliminated the need for human spies in space.

SIX

Eyes in Orbit

ALMOST AS SOON as the first United States satellites went into orbit, the American armed forces were making plans for reconnaissance satellites—the kind of spy system generals and admirals had dreamed about for decades. A satellite equipped with cameras could photograph huge areas of the world at once. It would fly so high as to be invulnerable to attack, whereas an aircraft, even such a high-altitude flyer as the U-2 spy plane, could be knocked down with rockets.* It would also be untroubled by picture-blurring influences such as an aircraft's engine vibration. Furthermore, an unmanned satellite would be much less expensive to launch than the manned observation platforms envisioned by von Braun and other rocket experts in the 1950s. So the U.S. proceeded with plans to launch unmanned reconnaissance satellites to spy on the Soviet Union.

The Soviets themselves had helped pave the way for this development, in a way they may not have realized at the time. When the first Sputnik went into orbit, it was more than a technological achievement; it was also a legal precedent. Until that time, there had been debate about whether or not it was lawful to send one's satellites over another nation's land area. But the Soviets sent up Sputnik regardless of whose territory it crossed and whose rights it might breach. And if the Russians took this action, they could hardly afford to blame anyone else for following their lead.

That legal roadblock out of the way, the Americans were left to concentrate on the technical problems facing their spy satellite program. And those problems seemed formidable.

What the U.S. wanted to do was take an airborne reconnaissance camera of the kind used in high-flying military spy planes and pack the whole system, lenses and film and all, into a volume less than that of an oil drum.

A technique called "folded optics" had been developed

* When the Russians downed an American U-2 in 1959, and captured the pilot, Francis Gary Powers, the resulting scandal rocked the Eisenhower administration to the point of ruining an imminent U.S.-Soviet summit meeting.

that allowed a camera with a focal length of several meters to be compressed into a package only slightly larger than a lunch pail. In a folded-optics system, the greater the focal length, the sharper and more revealing the photo.

Commercially available folded-optics telescopes provide some idea of what a military reconnaissance satellite can accomplish. From a quarter of a mile away, a camera equipped with such a telescope can photograph the pupils of a heron's eyes—a feat accomplished with a telescope scarcely longer than a man's hand. A much larger unit, capable of far better resolution, may be mounted in a satellite—even a small satellite.

Once the problem of shrinking the camera was solved, it was still uncertain how the images would be returned to earth. One method finally chosen was to send back the film in a special re-entry capsule.

The satellite was fitted with a developer capable of processing the photos while still in orbit. The film would be fed into a recovery module which, when full, would be ejected from the satellite and start its fiery drop through the atmosphere, after firing a set of retro-rockets to slow the capsule to suborbital speed. To protect the film from the heat of re-entry, the capsule was coated with a heat-resistant ceramic similar to that used in cooking-ware.

In theory, the quickest, safest, and most secure means of recovery would be to pluck the capsule out of midair with a "lasso" attached to an Air Force cargo plane. Its parachute lines snared by the lasso, the capsule could then be reeled aboard the plane and taken home. The Air Force equipped a few planes with just such a recovery apparatus and, in August, 1960, succeeded in the midair recovery of a Discoverer space capsule.

Discoverer continued testing early reconnaissance systems, eventually totaling thirty-eight flights. And even as this project was winding down, the U.S. was well along with its successor—SAMOS.

Short for Satellite and Missile Observation System, SAMOS was a bigger and more sophisticated version of Discoverer, with a reported camera focal length of about 1.23 meters. SAMOS was intended to keep watch on the Soviet Union's growing ICBM capability. Because it was no longer safe to send high-altitude spy planes over the U.S.S.R., satellites were the only practicable means of telling what kind of missiles the Soviets had under development, how many missiles there were, and where they were located.

The Russians had been blustering again about the supposed size of their ICBM arsenal, and Khrushchev had used the threat of massive ICBM power to force an East-West confrontation over Berlin. But U.S. military intelligence wondered if Russia's missile might was anywhere near as great as the Premier had boasted. SAMOS, the Pentagon hoped, would clear up the mystery.

The first successful SAMOS was launched from Vandenberg Air Force Base in California on January 31, 1961, and a new age in reconnaissance began. It was a time when the U.S. most needed the information satellites could provide. Khrushchev, on the strength of Russia's successes in missile and space technology, based his foreign policy on fear of Russia's missile power. SAMOS soon established, however, that Russia's missile force was only a tenth that of the United States—information which helped to keep the world from war in October, 1962, when Khrushchev tried to install intermediate range ballistic missiles (IRBMs) in Cuba for purposes of nuclear blackmail against the U.S. Eastern seaboard.

After the presence of the missiles was revealed by aircraft reconnaissance photos, Kennedy told the Russians to remove them. Khrushchev refused, and threatened the U.S. with nuclear war. Kennedy then called the Russian's bluff and placed a naval blockade around Cuba to halt the emplacement of more missiles. Khrushchev backed down.

Only a year later, Khrushchev was out of office, a victim in part of his own rambunctious policies. He was replaced by Alexei Kosygin, the new Premier, and Leonid Brezhnev, the new First Secretary of the Communist Party. Khrushchev had held both positions while in power. Now he was retired to a dacha near Moscow.

Under the new regime, the Soviets attempted to catch up with the U.S. in the field of spy satellite technology, and today the two nations are thought to be approximately equal in the orbital reconnaissance race.

The showpiece of the U.S. collection is Lockheed's 12-ton Big Bird, which orbits the earth at about 250 kilometers above sea level and is roughly the size of a railway tank car. Big Bird can swoop as low as 125 kilometers for close scrutiny of ground installations, and is believed to be equipped with several kinds of reconnaissance cameras.

As a rule, American reconnaissance satellites like Big Bird are built for long duty, and may stay up for months on end. Russia's spy satellites, on the other hand, are shorter-lived; they may stay in space for only a few days. This difference in satellite lifetimes might spell trouble for the U.S. in future conflicts, for the Soviets could probably send up new satellites as fast as U.S. anti-satellite weapons could knock them down; while the Americans, their satellites bigger and less numerous, would be hard pressed to keep up with the Russians'.

Other categories of United States and Soviet military satellites include:

Ferret satellites. Perhaps the most mystifying of all military reconnaissance satellites, ferrets are packed with electronic "ears" that pick up electromagnetic signals— radio and radar—from the ground some 280 kilometers below. These eavesdropping satellites were created for electronic espionage jobs that are difficult, dangerous, or impossible for ships and aircraft. Little is known about ferrets outside the Pentagon, but rumors have circulated in recent years that there are "superferrets" under development that can reportedly pick up virtually any radio signal louder than the faint "pop" from the spark of a cigarette lighter's flint.

Navigation satellites. Mariners have always needed to steer by the stars, and satellite technology has given them better "stars" to steer by. There is now a network of navigation satellites in geosynchronous orbit, their positions known precisely, so that ships at sea can lock onto them and determine their own locations down to a few centimeters in any direction. This degree of precision is, of course, useful for military operations such as rendezvous. (Navigation has come a long way from Queen Victoria's day, when the exact latitudes and longitudes of even such frequented ports as Rio de Janiero and San Francisco were not known.)

Soldiers on land also find navigation satellites useful. They enable an infantryman or tank commander to find out precisely where he is at any given moment—a reassuring thing to know, especially if one is fighting in unfamiliar territory. All one needs to get a fix on one's whereabouts is a small dish antenna that when pointed at the satellite says, in effect, "You are here." The Navy's navigation satellite program is known as NAVSTAR and is designed to encircle the world with "lighthouses" in the sky.

Ocean surveillance satellites. Though it is desirable to know exactly where one's own naval forces are at any given moment, it is even more imperative to know where the enemy is. Searching the oceans was a hopeless task only a few decades ago. United States commanders fighting the Japanese in the Pacific in World War II had to trust largely to luck (and the sharp eyes of scouts in aircraft) to locate the enemy; and then, of course, the scouts' best efforts might be foiled by a random fog bank or a blinding reflection of sunlight from the water.

Nowadays the ocean surveillance satellites watch the sea for anything that looks suspicious. If a Russian sub comes to the surface off Hawaii, an orbiting eye will monitor it until it submerges. And perhaps even after it dives, satellites will be able to track it. According to current reports of progress in laser research, ocean surveillance satellites may soon be able to peer into the deeps with the help of blue-green laser light whose wavelengths are thought to be good for penetrating the sea and locating subs on patrol.

Military communications satellites. During the Vietnam War, these satellites were used extensively to keep Washington posted on progress of fighting. Soviet military communications satellites include the Molniya, a series that probably doubles as communications and spy satellites.

Weather satellites. These too were used in Vietnam, to monitor weather conditions over targets in the North. Weather satellite cameras are not nearly as sensitive as those on photoreconnaissance satellites. In fact, the weather satellite cameras are desensitized to block out light from small point sources like cities.

The United States is increasingly dependent on satellites for defense-related operations. The Defense Meteorological Satellite Program gathers information about weather conditions. The Air Force Satellite Communications System carries messages to and from armed forces all over the world. The Satellite Data System handles communications from one polar region to another and helps the Air Force track satellites. Altogether, about three-fourths of all military electronic communications are sent through satellites. Soon that figure may be close to 100 percent. When General Alton D. Slay, commander of the Air Force Systems Command, addressed the annual meeting of the American Institute of Aeronautics and Astronautics in Washington, D.C., in 1979, he called satellites "the key element for worldwide long-haul communications," and added: "For the future, I see no diminishment of the utility of space for communications and data relay. I see only expansion."

Expanding our military satellite systems may do us little good, however, if the Soviet Union can knock them down. And it appears the Soviets can do exactly that.

SEVEN

The Shooting Gallery

T H E 1967 International Outer Space Treaty, intended to slow the military buildup in orbit, forbade placement of weapons on other celestial bodies—the moon, asteroids, and so forth—and banned weapons of mass destruction, such as atomic bombs, from being put in earth orbit. When the United States and the Soviet Union signed the treaty, it appeared to remove the threat of spaceborne H-bombs and put a cap of sorts on the international spacewar race.

The language of the treaty, however, still allowed military activity in space, such as a series of tests that the Soviet Union started conducting on September 17, 1966. The Russians, as part of their seemingly endless series of Cosmos satellites, at that time began launching space vehicles with odd trajectories. Instead of going into orbit, these vehicles arced up far above the altitude normal for a reconnaissance satellite, and then fell back to earth without completing a full circuit of the globe.

It looked rather as if the Russians were engaged in some kind of target practice. That, in fact, was what was happening.

At a hastily called press conference in November, 1967, Defense Secretary Robert McNamara revealed that the Russians were testing a Fractional Orbital Bombardment System, or FOBS. It was designed, he said, to drop a nuclear bomb on a target from outer space within a fraction of a single orbit; hence the system's name.

The FOBS tests followed the same basic pattern as the V-2 attacks on London during the Second World War, only the FOBS flew much higher and farther. FOBS soared to an altitude of 1,120 kilometers and traveled perhaps a fourth of the way around the world before falling on target.

The goal of FOBS was to circumvent America's first line of defense against Soviet ICBM assault, the Distant Early Warning (DEW) line, a string of radar outposts along the northern fringe of the continent. Like giant golf balls perched on giant tees, the radar domes rose out of the tundra all the way from Greenland to the tip of the Aleutian Island chain. Their foundations reached into permanently frozen soil.

If war broke out, the DEW line was expected to spot Soviet missiles roaring over the icecap at an altitude of about 140 kilometers.* As soon as the incoming ICBMs showed up on the radar screens in the far north, warnings would be flashed to the lower forty-nine states to brace for a nuclear attack. The DEW line would give America roughly fifteen minutes' notice before the bombs struck.

The DEW radars were aimed along the horizon and could spot an ordinary ballistic missile on its way across the Pole. However, the Russians could slip past the radars by putting their missile into a very high trajectory where the DEW radars would not detect it. A FOBS bomb would be picked up on radar only when it was about 700 kilometers away from the target—practically on the doorstep, considering how fast the missile traveled—and by that time there would be only three minutes' warning before the warhead went off.

In his press conference, McNamara suggested that FOBS was designed mainly for use against relatively "soft" targets such as Strategic Air Command bases. There heavy bombers would be waiting for the scramble order to attack the U.S.S.R. Unprotected on the field, the bombers would be sitting ducks for a sneak FOBS attack. The crews would be hard pressed to get their planes airborne before the FOBS missile reached them and blasted them into ionized

* Contrary to popular belief, Soviet and American land-based ICBMs are not aimed across the Atlantic and Pacific oceans at their targets. They are aimed instead along a Great Circle route designed to take them to their destinations as quickly as possible; and most of those routes lie across the Arctic. A Great Circle is the intersection of the earth's surface with a hypothetical plane running through the center of the planet.

gases. And even if the planes did get off the ground before detonation, the terrific shock wave from the explosion would probably be enough to knock them out of the sky.*

McNamara tried to sound reassuring. He told the press that the FOBS was markedly less accurate than a standard ICBM. That seemed small comfort, however, because a soft target such as an air base need not be hit directly to be put out of commission.

In addition, McNamara said that the United States had a new over-the-horizon radar which was not limited in range by the curvature of the earth. This radar, according to the Defense Department, could spot a FOBS missile taking off from the Soviet Union within seconds after the booster left the launch pad.

The over-the-horizon radar would increase America's warning time to perhaps half an hour. That would give Washington time to activate its planned anti-ballistic missile (ABM) system against the FOBS attack.

The ABM was a two-part defense. The first part consisted of a three-stage missile called Spartan. It would be launched against enemy missiles to destroy them at a safe distance from the target—"safe" meaning that persons on the ground would be spared fallout from the explosion of Spartan's warhead. The Spartan would either knock the

* Another, even softer target for the FOBS was North America's telecommunications system. A high-yield nuclear warhead set off several hundred kilometers high would generate a titanic "electromagnetic pulse"(EMP)—a burst of "noise" that would blind and deafen much of the U.S. and Canadian military telecommunications setup, as well as Civil Defense broadcasting. A mighty EMP would put North America at a grave disadvantage when it came to striking back against a Soviet FOBS assault.

approaching missiles off course, blow them to pieces, or fry their sensitive guidance systems with radiation from Spartan's fireball.

Any ICBMs that slipped past Spartan were to be destroyed by a much smaller but faster missile called Sprint. Sprint would shoot up at a phenomenal rate of climb and explode in the path of the incoming missiles, knocking them down or incinerating them in midair. Some fallout from Sprint was bound to drift down onto nearby cities. Sprint was the defense of last resort.

By 1967, however, America's planned ABM system was starting to look impractical. Scientists were all but unanimous in pointing out that no ABM system could halt a determined Russian attack. There were just too many ways to outfox the ABM system. The simplest way was to throw so many warheads at the target that one or more would be sure to get through. Only one was needed to destroy a city.

A more refined plan was to detonate a nuclear warhead just short of the target, creating a fireball that would block the "vision" of the radars used to guide the interceptor missiles to their prey. The Russian missiles would then slip in behind and over the top of the fireball, and drop straight down on the unprotected target.

It looked, therefore, as if the ABM system, even if built and deployed, might not prove much safeguard against a FOBS attack. And McNamara suggested that the Soviet Union might have an operational FOBS system ready in 1968, long before any American ABM was ready.

All at once the old nightmares about nuclear bombs in space returned to haunt the U.S.

The *New York Times* warned that a FOBS might escalate the arms race. If one side had the FOBS, the other side would surely feel compelled to build it too. The *New Republic,* from a legal point of view, hinted that Russia's new bomb delivery system might violate the International Outer Space Treaty, which forbade the signatories to station weapons of mass destruction *in orbit.* But FOBS was not really "stationed" in space. Rather, it would merely be passing through space in the course of completing its mission.

If the FOBS had been an orbiting spacecraft like some envisioned years earlier, loaded with a nuclear warhead ready to be called down on a bombing run, then the 1967 treaty most assuredly would have been broken by the U.S.S.R. But since the FOBS would accomplish its deadly work in much less than a single orbit, popping only briefly into space en route, it might be argued that FOBS was indeed permitted under the terms of the agreement. FOBS was just one of the first signs that legal efforts to curb the space race might end up permitting more military operations in orbit than they banned.

The FOBS tests continued until 1971, when the United States and Russia edged cautiously into the era of détente. Nineteen seventy-one also marked the temporary end of another series of tests that had U.S. defense officials worried.

Starting in 1968, the Soviets had begun sending up Cosmos satellites on missions that looked suspiciously like the beginnings of an anti-satellite system somewhat like DOD's old Saint. The missions proceeded as follows: one satellite would be sent into orbit; then another would follow it, pass within a few kilometers of it, and, a few minutes later, explode.

For American spacecraft, the implications were frightening. If the Soviets could intercept their own spacecraft in orbit, then they could do the same to ours.

No longer were reconnaissance satellites such as SAMOS invulnerable to attack. And the loss of those satellites might prove disastrous to the United States in any future conflict. Communications satellites might be struck down as well, though this was less of a threat, because many of the military comsats were placed in high geosynchronous orbits, more than 30,000 kilometers out, where the low-flying Russian killer satellites could not reach.

Again the suspicion arose that the Russians were playing fast and loose with the International Outer Space Treaty, but in fact Russia's ASAT missions did not specifically fall into any of the categories of military operations covered under the pact. An anti-satellite weapon, after all, is not a weapon of mass annihilation. It has nothing like the destructive power of a nuclear bomb. It is directed only against a few small targets in space. And it is not stationed on the moon or any other celestial body. Once again, the Russians had wriggled through a loophole in the treaty.

But at the same time, they had, as with Sputnik, set another legal precedent. Now the Russians could hardly object to the U.S. developing its own ASAT weapons.

During the warm years in U.S.-Soviet relations, the Russians refrained from testing killer satellites in space. That did not mean, however, that the Soviet ASAT effort was halted entirely.

On the ground, the Russians were still busy, reportedly conducting dry runs to refine their ground operations for future ASAT tests. The more time the Soviets could shave off their preliminary work on each ASAT mission, the faster the attacks could be carried out—and the less time America would have to realize its satellites were under assault.

Ironically, the intended targets of these attacks, U.S. reconnaissance satellites, were peering down at Russia all the while American and Soviet leaders were toasting their newfound friendship. What the satellites photographed made détente look tenuous indeed, if not a total illusion.

Russian ground crews could be seen moving rockets tipped with killer satellites out of their hangars at the Tyuratam launch facility on the east side of the Aral Sea, in Soviet Central Asia. The Russians were able to set up the rockets on their launch pads, fuel them, and prepare for launch in less than ninety minutes.

Tyuratam is one of three Soviet launch facilities, and is probably the most widely known, for it is used to send up the Soviets' heavily publicized civilian space missions. Tyuratam is also known as Baikonur, which was a clumsy effort to mislead Western intelligence about the true location of the rocket base. Baikonur is a mining town not far from Tyuratam.

About a thousand kilometers west of Tyuratam is a much smaller satellite launch facility, in Kapustin Yar. This base is equivalent to the old NASA launch site at Wallops Island, Virginia, which was mainly for upper-atmosphere studies using relatively small rockets, and was closed in the late 1970s as an economy measure.

The third and most secret of Russia's space bases is Plesetsk, between Moscow and the northern port of Archangel. This is the Soviet Union's military spaceport and is equivalent to the Vandenberg Air Force Base, starting point for many U.S. military space missions.

Plesetsk is one of the world's biggest launch facilities, if not *the* biggest, and one of the busiest as well. In 1973, for example, a satellite went into orbit from Plesetsk on an average of every four and a half days. The spaceport at Plesetsk is served by its own airfield and highway system, and by the Moscow-Archangel railway.

About 1965, U.S. satellite photos showed construction in the Plesetsk area, and in 1966 satellites began taking off from the new military spaceport. At the time, no one outside the intelligence community in the U.S. was sure what was going on at Plesetsk, but soon the secret was out.

Among the first Western civilians to detect the existence of Russia's new space base were students at the Kettering Grammar School in England. Using secondhand radio equipment, the British youths tracked Soviet satellites and plotted their orbital paths. However, when they plotted the orbit of the first satellite launched from Plesetsk—Cosmos 112—they found something odd about it. The orbit did not originate from Tyuratam in Central Asia, but rather from somewhere far to the north. Later launches followed the same pattern, and the students deduced that there must be a new launch facility in the northwestern U.S.S.R.

Though the Soviets denied the existence of any such facility in that part of the country, they were caught in their lie a few years later, when the giant base appeared in photos taken by a civilian U.S. LANDSAT (Earth Resources Technology Satellite) launched in 1973. The LANDSAT shots showed a large military airfield with scores of launch pads. Rather than call attention to its military space center, the U.S.S.R. let the discovery of Plesetsk pass without official comment. To this day, Moscow refuses to acknowledge the existence of the base.

Most of the military satellites launched from Plesetsk are part of the all-purpose Cosmos series, a designation that can cover anything from a FOBS missile to a biological research satellite full of fertile quail eggs. The cover of the Cosmos name, as later chapters will show, has come in handy more than once for the Russians when they wished to put security wraps on a satellite or try to hide a serious blunder or accident.

The Russian ASAT missions have all been conducted as Cosmos flights, and they resumed in earnest after détente began to deteriorate in the mid-1970s. So far, the Russians have scored more than a dozen successful "kills" on targets in orbit.

Especially worrisome to the U.S. is a recent series of ASAT tests that are evidently meant to develop a one-orbit intercept capability: the killer satellite would zoom up and destroy an American satellite within a single orbit of the attack spacecraft. Such a quick shot would leave the U.S. little time to realize that one of its spacecraft was being assailed. To avoid being caught by surprise, the United States is identifying "threat windows"—times when our satellites might be in particular danger from Russia's ASAT missiles.

To keep close watch on anything resembling an ASAT attack, the Pentagon has improved its space-surveillance network. The brain of this network is a new Space Defense Operations Center, in Colorado Springs, Colorado, located at the NORAD center under Cheyenne Mountain. Here, for the first time, DOD has a single site at which to monitor the status of all spacecraft connected with the nation's defenses.

This locus is being tied in with all American military satellites, as well as civilian space systems, such as weather and communications satellites, that might be targets for Russian ASAT weapons. Computers at Cheyenne Mountain receive data from all NORAD radar sensors (including an installation in North Dakota that can allegedly spot a low-orbiting object the size of a cigarette lighter) and can keep track of 8,000 space vehicles at once. At any given time, there are only about 1,000–2,000 satellites in orbit, but there are also many other items of "junk"—spent boosters, for example—that must be tracked so that no one will confuse them with satellites and draw the wrong conclusions when they re-enter the atmosphere or collide with one another. (Collisions between satellites, once an almost unthinkable occurrence, are becoming such a serious hazard that before all U.S. Apollo and Skylab missions, the NORAD computers had to run checks on everything in orbit to make sure nothing would run into the spaceships. At least two U.S. communications and scientific satellites are thought to have been damaged or destroyed by impact with other pieces of hardware in space.)

Radars are not the only surveillance systems scanning space for DOD. While camera-carrying satellites are turning their lenses toward the earth, ground-based cameras are looking back at the spacecraft.

Since about 1955, the U.S. has had a set of powerful Baker-Nunn satellite surveillance cameras scattered about the globe to monitor satellites. The telescopes linked to these cameras used a special drive unit to correct for the motion of the ground relative to the stars. So the star field stayed motionless in the background, while satellites showed up as light streaks against the darkness of space.

The Baker-Nunns are due to be replaced soon with an awe-inspiring new space-surveillance apparatus known as GEODSS—Ground-Based Electro-Optical Deep-Space Surveillance System. What makes this system different from the Baker-Nunn is the "electro" in its name.

Rather than simply gathering light and focusing it on film, as a camera does, GEODSS will convert light into electrical impulses that will then be fed into a pair of monitors. This will be quicker than the Baker-Nunn system, which required time for the developing of film after exposure. GEODSS will also be sensitive to even faintly reflective objects in orbit. Under optimum conditions, it is supposed to have a 95 percent chance of detecting anything that orbits across its field of view. There will be five of these colossal electronic visual aids spaced roughly equidistant around the globe. The Baker-Nunns are reportedly capable of seeing something the size of a watermelon in geosynchronous orbit 32,000 kilometers high. GEODSS is expected to do much better. If something is out there, chances are the blue-suited men under Cheyenne Mountain will know about it.

The Space Defense Operations Center will be the control point for any future U.S. ASAT operations. Afraid of falling far behind the Soviets, the United States has been working feverishly since the mid-1970s to build its own anti-satellite capability.

One of the results is a small homing missile about the size of a large can of juice. Equipped with a rocket propulsion system and infrared sensors for seeking its target, the U.S. missile will be launched either from the ground or from an F-15 fighter aircraft flying near the boundary of outer space. This system is expected to be more flexible than the Soviet killer satellites—and cheaper, because it could be launched with smaller boosters than the big SS-9s the Russians use to launch their killer satellites.

The Pentagon is quiet on the topic of which Soviet satellites would be prime targets for a U.S. ASAT mission, but Russia's ocean surveillance satellites are surely near the top of the roster, for much of any future war will be fought on and under the sea. Communications satellites will also be high on the hit list, but here the Soviets are less vulnerable than we are. While the U.S. has phased out many of its cable communication systems in favor of satellites, the Russians have kept most of their non-satellite communications setup on earth working. This "redundancy" (the systems engineer's expression for having two ways to do any given job) means the Russians might be able to fall back on their old, less sophisticated, but also less fragile communications channels if their military relay satellites were knocked out. The Americans, by contrast, would be stuck without their spacecraft. That is one reason the U.S. is trying so desperately to match Russia's ASAT capability as quickly as possible. If ASAT warfare broke out tomorrow, Moscow's armed forces would probably have a considerable edge on the U.S. because of our extreme reliance on spaceborne communications links.

Dr. Wernher von Braun. Von Braun and many of his colleagues from Nazi Germany were instrumental in starting America's military space program.

Gemini I spacecraft. Gemini, a two-man space capsule, was to serve in part as a military spaceship. Plans called for the Gemini to ferry astronauts to and from a military space station called the Manned Orbiting Laboratory (MOL) in the 1960s, but MOL was canceled for budgetary reasons before the station could be orbited.

Astronauts Neil Armstrong and Elliot M. See practicing "water egress" training exercises from a Gemini spacecraft. The Gemini version shown here is the civilian one, with hatches in the side of the capsule. The military Gemini, known as Blue Gemini or Gemini X, would have had a hatch cut in its rearward-facing heat shield, so that astronauts could crawl directly from the capsule to the MOL after docking with it, avoiding the need for a hazardous "spacewalk" to the MOL from Gemini.

An artist's concept of the Pegasus satellite in orbit. Launched in the 1960s to study the possible danger from micrometeorites striking a spacecraft's hull, Pegasus was about the size of many modern spy satellites, or roughly as big as a tank truck.

The Apollo 13 service module. Accidents happen on space missions, as this photo of the Apollo 13 service module shows. The failure of a tank on board the Apollo demolished much of the service module and came close to killing the astronauts on that mission. There is a chilling possibility that accidents involving future military spacecraft might be mistaken for acts of war, and trigger armed conflict in space and here on earth.

An early concept of the NASA space shuttle. The shuttle is the kind of spacecraft that the Defense Department has wanted since the U.S. space program began: a reusable manned spaceship suitable for a wide variety of military missions.

The modern NASA space shuttle. Though nominally a civilian project, the shuttle is in large part a military undertaking. The shuttle was designed to Pentagon specifications and was saved from cancellation, during the Carter administration, because the armed forces wanted the shuttle for their own space effort.

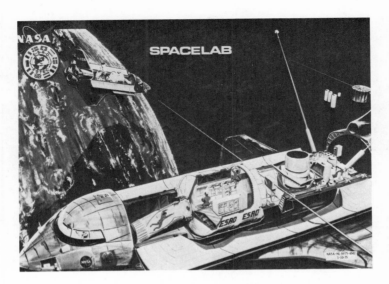

Spacelab (artist's conception). A joint U.S.–European undertaking. Spacelab is similar in many ways to the old MOL.

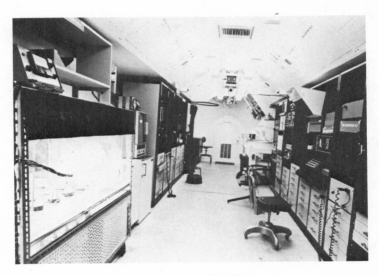

Spacelab, interior view.

Soviet Soyuz spacecraft in earth orbit. Roughly comparable to the U.S. Gemini, Soyuz has been used to carry cosmonauts to and from Russia's quasi-military Salyut space stations.

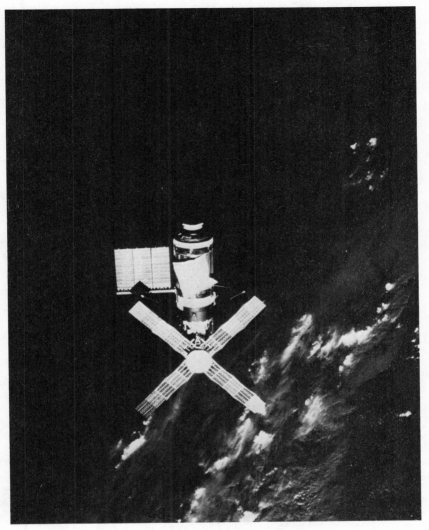

View of Skylab space station. U.S. spacecraft designers were thinking of a military space platform much like Skylab, for reconnaissance and other Defense Department uses, as early as the Eisenhower administration.

Artist's conception of a solar power satellite. The technology developed to build such large non-military structures in space might also be used to build military space stations on a grandiose scale.

EIGHT

"[Deleted]"

H. G. WELLS, in his 1898 novel *The War of the Worlds,* described a "heat ray" that invading Martians used to defeat the earth's armed forces. The Martians carried the heat ray generators in mysterious boxes on huge tripod walking machines, which they used to devastate the English countryside. Wells was vague about how the beam was propagated, but it was powerful enough to cut through armor plate. At one point, the Martians turned the ray on a naval vessel, touched off the ship's magazines, and blew it out of the water.

In the face of the Martians' advanced technology, humans could only retreat. They were ultimately saved, however, in one of the classic surprise endings in literature. Lacking all resistance to terrestrial microbes, the Martians fell prey to the bacteria and died of massive infections, surrounded by their now useless weapons.

Wells's heat ray inspired many imitations. Science fiction writers of the 1920s and 1930s would often enliven their tales with "death ray" battles, though few authors seemed to have any idea how such a weapon might work. An exception was Raymond Z. Gallun, whose 1929 story "The Crystal Ray," published in a U.S. pulp magazine, featured a beam of energy generated by a special crystal and used in aerial warfare.

The heat rays and death rays were joined by another kind of ray in the 1920s and 1930s—the disintegrator ray. It worked by somehow cutting through the bonds of force that held matter together. Depending on who was writing about it, the disintegrator ray could either make solid objects disappear into vapor or could cut through spaceship hulls as neatly as a surgeon's knife through flesh.

The heat/death/disintegrator rays seemed likely to remain a dream until about 1960, when researchers in the United States and the U.S.S.R., working independently, developed the first laser.

"Laser" is short for "light amplification by stimulated emission of radiation." Though a complex machine, it has a basically simple job. It brings order to light.

Light is made up of waves. But this is something of an oversimplification. According to the modern theory of light, it is a "wave-particle phenomenon," meaning it behaves as if made up of waves or discrete particles, depending on one's point of view. For our purposes, however, the wave model will serve.

The wavelength, designated by the Greek letter lambda (λ), is the distance from the crest of one wave to the crest of the next, thus:

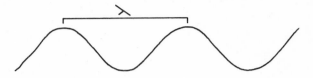

Wavelength determines, among other things, the color of light that we perceive. Red light has much longer waves than blue light does—a fact that explains why the sun looks redder at sunrise and sunset than at midday. In the early morning and late afternoon hours, the sunlight passes through a longer reach of atmosphere than it does in the middle of the day. Short wavelengths are thus filtered out, leaving only the red-orange part of the spectrum to reach our eyes.

Unless filtered or otherwise modified, ordinary light like that from a flashlight contains many different wavelengths, short and long and inbetween:

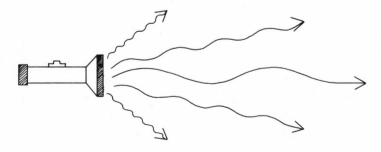

This mixture of wavelengths is "non-coherent" light. The opposite of that, of course, is "coherent" light—and this is where the laser comes in.

A laser replaces the bewildering range of lambdas in ordinary light with a single wavelength. But it does much more than merely make the wavelength uniform. The laser lines up the waves in neat rows, so to speak, with crests and troughs all synchronized:

Now the waves are no longer going their separate ways. They are "hanging together," one might say. They are coherent.

Besides its attractive symmetry, there are thousands of things that make coherent light special. These highly ordered light waves can do things that are impossible for an ordinary light beam, simply because they are ordered in this particular way.

For one thing, laser light can concentrate enough energy on solid steel to burn right through it. For example, imagine a seawall with millions of individual waves beating against it every day. As long as the waves arrive in many different sizes and periods, they will dissipate their energy harmlessly and do the wall no damage at all. They cannot concentrate enough of their energy at one time to break the resistant steel and concrete, because the waves are non-coherent.

Now suppose a tropical storm passes just offshore, and

its fierce cyclonic winds raise a strong periodic swell. The swell will reach shore as a long string of big, energetic waves with roughly even spacing. Hitting the seawall one at a time, like battering rams, the waves will concentrate their energy in sharp, repeated blows that can make short work of even the sturdiest seawall.

"Coherent" sea waves like these have been known to knock down lighthouses and rip holes in barrier islands. In this example, the storm acts much as a laser does. It organizes and regularizes waves so that they pack more punch.

Coherent light has turned out to have amazing properties that are only beginning to be applied to space technology. One of the most intriguing proposals is a laser propulsion system for spacecraft. Science fiction illustrators have long depicted spaceships rising into orbit on powerful light beams, and lasers may accomplish just that in the near future.

This is how a laser launch would proceed. A spaceship would be loaded with propellant in preparation for its trip into space. Then a laser beam would be focused on the spaceship from another point on the ground. The beam would enter the spacecraft through a port in the side and would be reflected by an internal mirror into an absorption chamber, where the laser light would vaporize the propellant to provide thrust for lift-off. The laser would remain focused on the vehicle during its ascent, to keep the engine running. The beam would, of course, have to be extremely powerful, and subterranean H-bombs have been suggested as a possible power source.

Another intriguing plan is to turn rockets into electrical power plants by means of an exotic branch of physics called "magnetohydrodynamics," or MHD for short. MHD deals with the electrical properties of plasmas, the so-called fourth state of matter: something like a superhot gas. A candle flame is an example of a plasma. So is the fiery exhaust of a rocket, and that is why a rocket could be turned into an MHD generator to make electricity.

Set the rocket on its side and ignite the engine. Then "season" the exhaust with metal salts that will ionize the plasma—that is, make it a conductor of electricity. Finally, direct that electrically conductive flame through a powerful magnetic field. The result will be an electric current. (This MHD effect has been known for more than a century.) Draw off the electricity through electrodes set in the walls of the tunnel through which the hot exhaust is passing, and then feed the power to the laser. This method of generation would be highly efficient (in theory, an MHD unit could approach 60 percent efficiency, compared to perhaps 5 or 10 percent for a coal- or oil-fired power plant). MHD generators could be fueled with nuclear bombs, too. A single explosion would generate tremendous amounts of plasma for making electricity: enough to send whole fleets of laser-driven spaceships into orbit.

Rather than shooting energy into space, lasers could instead be used to transfer solar energy from outer space to earth. One plan calls for the building of giant solar collectors in orbit to gather sunlight and use it to "pump" a laser, the engineers' expression for delivering energy to the laser in order to generate a beam. The laser would convert the sunlight into coherent light to be fired to receiving stations on the ground, where it would be turned back into electricity and fed into power grids to light cities. There is practically no limit to the size of the collector in this plan. Solar satellites with the area of small nations have been proposed. And they could work twenty-four hours a day, seven days a week, providing a tremendous, nonstop, pollution-free source of power for the world. To avoid interference from clouds and haze on the ground, receiving stations on earth could be built in the desert.

The peaceful uses of lasers in space have been headline material for more than a decade. Their military side was largely ignored in the media, however, until the last few weeks of 1975, when reports surfaced in the press about an alleged laser attack on two U.S. satellites over Asia.

These satellites were stationed in geosynchronous orbit, scanning the U.S.S.R. for the telltale bursts of IR energy from rocket launches or nuclear tests. Suddenly, the news stories said, the satellites were "blinded" by intense radiation. The reports claimed that the satellites had been neutralized by a powerful laser beam fired at them from the ground.

The Pentagon denied that report was true. The satellites, DOD claimed, had merely been "dazzled" by the light from large fires along natural-gas pipelines in the U.S.S.R.

Cartoonists had a field day with the "zapping" of the U.S. satellites. Bill Mauldin, of the Chicago *Sun-Times,* depicted a malicious-looking bear popping off two American spacecraft with a laser "rifle."

Following a decade of relative silence about the military uses of space systems, war in space was suddenly a hot topic again in the media. Rumors began to circulate that the U.S. had never given up its old military space projects, such as Saint and Early Spring, but had merely filed them away for future use. And Pentagon officials dropped hints about U.S. interest in using laser ASAT technology to establish "anti-space" for Soviet satellites—that is, to knock Russian spacecraft out of the sky with laser beams. Defense Department public statements contained veiled and tantalizing references to "deterrent forces" that might be placed in outer space sometime in the future. These hints encouraged the media to exercise their imagination, and soon the press was full of stories of projected weapons systems that sounded much like the assorted beams of science fiction's pulp-magazine days. A surprising number of these reports said they were "based on Defense Department studies," or words to that effect.

Actually, most official statements about ASAT lasers were less than revealing, as *Astronautics and Aeronautics* suggested in a 1977 article about the new anti-satellite fever. The magazine quoted from a Navy document submitted to the Senate Appropriations Committee: "Using [deleted] devices with power levels of [deleted], these types of kills could be effected at ranges out to [deleted]."

Dr. Malcom Currie, Director of Defense Research and Engineering, was quoted in 1976 as saying the U.S. was "investigating the vulnerability of our satellites to radiation from lasers and . . . examining techniques for reducing the effects of such radiation"; that is, "hardening" satellites against laser-beam assault.

Currie added, in a policy statement for that year, that "the Soviets have a comprehensive program in laser research and development, and . . . are leading us in some areas." And the Defense Advanced Research Projects Agency (DARPA) said in a 1976 policy statement that current laser research "could lead to a device whose size and weight would enable it to be used in space."*

These cautiously worded statements said essentially what everyone already knew. Lasers were becoming tools of war, on the ground as well as in space.

Reports of laser weapons had been appearing in the media for years before the laser ASAT flap of the mid-1970s. The trade journal *Laser Focus,* for example, had reported in 1972 that a laser had been used to ignite a wooden target some three kilometers distant. This particular project was code-named Eighth Card and was said to be based at Kirtland Air Force Base, near Albuquerque, New Mexico. Evidently a good tracking system was hooked up to the Eighth Card laser, because the weekly *Science News* reported that the beam could burn a hole through something the size of an ace of clubs, waved at the tip of a pole more than a kilometer away.

It appeared that a "superscientific struggle" for laser-weapon supremacy, as *Jane's Weapons Systems* described it, was in the making between the United States and the Soviet Union in the mid-1970s. But what kinds of lasers were under development, and how effective would they be? Shooting holes in playing cards is one thing; shooting down missiles, aircraft, and satellites in actual combat is

---

* Size and weight had been strong arguments against the feasibility of space-based laser weapons until the advent of the NASA space shuttle. With its huge cargo capacity, however, the shuttle stood to make possible the orbiting of 30-ton payloads—and suddenly heft seemed less an obstacle to hauling heavy laser hardware into orbit. Previously, engineers had joked that if one had a laser capable of doing enemy forces any harm, one wouldn't even have to fire it: just drop it on the enemy's capital city.

something else again. Was talk of ASAT lasers merely so much speculation, or did the military seriously expect to deploy offensive and defensive laser weapons in the near future?

Congress was told in 1976 that a high-energy chemical laser was a good bet for application in space "because of its higher mass efficiency and ability to produce laser power [without] a large electrical power supply." The chemical laser—unlike early models, which consisted of a synthetic ruby crystal hooked up to an electric power source—draws its energy from fierce chemical reactions that liberate light at certain wavelengths. So if the chemicals could be lifted into orbit, they could be stored for mixing whenever needed to create a laser beam.

More specific was a 1978 report in *Aviation Week,* which said that a deuterium fluoride laser seemed the most likely candidate for anti-missile uses. The laser would emit infrared radiation and damage its target primarily through thermal effects—in other words, "cook" it.

Also discussed in the article was a promising new kind of laser called an Excimer. The name was a fusion of "excited" and "dimer," the expression for a chemical formed by the combination of two molecules of a simpler compound. The Excimer laser generates a beam by reactions involving a compound that exists only when one or more of its components is in an excited—that is, high-energy—condition. The Excimer would shoot out visible light, as opposed to the invisible IR beam from a deuterium fluoride laser.

By late 1978, the Air Force was deeply involved in laser weapons research and development. Worried about the growing number of Soviet ASAT tests—sixteen of them

between the autumns of 1968 and 1978, with ten successful interceptions—the Air Force, *Aviation Week* said,
was working intensively on ASAT weapons to meet the Soviet threat in orbit.

One proposed plan was to build a ground-based laser
to fire up at enemy spacecraft. The laser would be set up
in the desert or some other location, such as a mountain-
top, where weather conditions would not interfere with
firing. Even a short sprint through the atmosphere, however, could weaken the beam, because of a phenomenon
called "thermal blooming," in which the laser beam heats
up the air molecules slightly as it passes, draining away
energy from the beam itself and making it spread out and
lose its tight, cohesive character. Such "blooming" makes
the beam less effective as a weapon. Ideally, there should
be no atmosphere at all between the laser and its target,
and the only place that condition prevails is in outer
space.

So the Pentagon has been investigating the possibility
of stationing laser battle stations in orbit. One such project is known as Talon Gold. A scaled-down version of
Darth Vader's Death Star in *Star Wars*, Talon Gold would
use a multimillion-watt laser to destroy enemy missiles
and satellites.

There is even talk of using lasers to destroy ships at
sea—including, possibly, the oil tankers on which the
West depends for its economic survival. If the Russians
could blow up supertankers with a well-directed laser
beam, they might be able to impose their will on virtually
all the world's non-Communist oil importers, including
the United States, Japan, and Western Europe.

Supertankers would make excellent targets. Some are

115   "[Deleted]")

almost half a kilometer long. Moreover, at certain times they are full of highly combustible vapors that turn the ships into floating bombs, just waiting for the tiniest hint of ignition to make them explode. The thought of tankers going up in fireballs all over the world's oceans is enough to make anyone tremble. Even one sunken tanker could mean disaster for the West.

Imagine a Soviet superlaser in orbit, zeroing in on a huge tanker in the narrow Strait of Hormuz, through which all the giant ships must pass on their way to and from the rich Persian Gulf oil fields. A quick pulse from the laser, and the tanker disappears in a cloud of flame. Its hulk settles to the bottom and blocks the channel to further traffic. Immediately the oil fields are closed off. Clearing the channel will take weeks or months—a sunken wreck the size of the Empire State Building is no easy thing to remove—and in the meantime the United States and its allies will be starving for oil. A single laser salvo brings the West to its knees.

With such scary scenarios in mind, the U.S. government has recently increased its funding for military laser weapons. And, according to reports of the laser tests, the money has bought some impressive results.

The Navy used an experimental laser of the deuterium fluoride kind to blast a tethered helicopter out of the sky in 1980. As part of its Sea Lite program, the Navy intends to beef up and refine its lasers for use against more difficult targets, such as missiles.

The Air Force also makes claim to heartening results from its laser experiments. In January, 1981, a laser test on board an NKC-135 jet prompted Air Force Secretary Hans Mark to call the results a "significant milestone."

Although it was reported that no laser beam was actually fired from the airplane, Mark said he thought the U.S. could "now think of shooting down the other fellow's missiles without using nuclear warheads."

A few weeks after that test, the Boston *Globe* printed a story on a Pentagon study calling for an accelerated space-laser program, aimed at putting an "armada" of weapons platforms like Talon Gold in orbit in the near future. These stations, the report said, would number about a hundred and "could cope with a simultaneous launch of one thousand ICBMs." The highlight of the article was a report on a new energy source for such orbiting weapons: atomic bombs.

A small nuclear explosion would be used to generate a brief but intense pulse of x-rays. The x-ray beam would be aimed by special telescopes and could, in theory, destroy hundreds of missiles while they were on their way to their targets. Most likely these ultra-powerful x-ray lasers would be supplemented by less impressive—but still formidable—chemical lasers stationed in space.

Not everyone, however, sees in lasers much promise for warfare.

In a 1980 study on the subject, Dr. Kosta Tsipis, of the Massachusetts Institute of Technology, a widely respected authority on lasers and their military uses, took a dim view of their potential as orbiting weapons. Any effective kind of laser weapons system, says Tsipis, would have to be monumental in size, and it would take several hundred such satellites to cover the world adequately. Each individual laser platform would be a major project, requiring three or four launches of a U.S. space shuttle merely to carry the parts into orbit.

Eight hundred to a thousand shuttle flights would be too much work for America's shuttle fleet to accomplish in its projected lifetime. The most optimistic estimates put the maximum number of shuttle missions in the 1980s at only about five hundred.

Also, a single laser "battle station" would have a staggering price tag: probably somewhere close to $12 billion. Even a relatively skimpy armada of a hundred such stations would come close to exhausting the gross national product of the United States for a single year; and, as the fate of MOL demonstrated, the fiscal resources of DOD, while great, are still finite. There are some security measures that even Uncle Sam cannot afford.

Another drawback to orbiting superlasers would be their extraordinary power demands. A ground-based laser could be tied to a colossal power source, but an orbiting laser must carry its own energy supply—a highly expensive proposition, in view of what it would cost to lift extra fuel tanks or nuclear reactors into orbit.

Perhaps the biggest problem with lasers as weapons is that they could probably be defeated by cheap and simple countermeasures. A mirror would do. Coat a satellite with silver and the laser beam will bounce off harmlessly. A reflective "parasol" like those used by photographers would do just as well. And if the satellite is attacked by lasers from two quarters at once, simply deploy another parasol. Such a defense would add little to the weight of a satellite and, for a few thousand dollars of added expense, might foil the best efforts of a $12 billion laser. If no parasol is available, and the satellite's skin is less than mirror-shiny, then it would be easy to spin the spacecraft so that the laser beam could not impinge on one spot long enough to burn it through.

Under optimum conditions, a spaceborne laser might make short work of a ship at sea or soldiers on the ground. But how often are optimum conditions found for *any* weapon?

The laser would have to have a clear shot through the atmosphere to the target. That means no clouds or fog that would intercept the beam. Even a faint haze could scatter the laser light to the point of inefficacy.

Except for the deserts, the earth's surface spends much of its time under cloud cover of one kind or another. A laser could be stopped cold by a high layer of cirrostratus—or, for that matter, by the smog over New York City or Los Angeles. If clouds were missing on a given day, and a laser assault from space seemed imminent, it would be no great trouble to generate smoke. Buildings could be aluminized in a matter of days and would be all but immune from laser damage thereafter. So could missiles. Persons in the open could carry a lightweight reflective shield of some kind. Mirror siding for a building might cost $5,000. A personal reflector—something like an umbrella—might cost as little as five dollars. A single laser salvo might consume $20,000 in fuel, and could be parried for little more than pocket money.

Even if lasers are unpromising weapons, there may be a bright future for them in space as communications systems; a laser beam could carry tremendous amounts of information and would be difficult for an enemy to jam. Radio communications, by contrast, may be blocked by broadcasting interference on the same frequency as the desired transmission. Radio is also easy to overhear, if one has the frequency of the enemy's chatter, because radio waves ripple out in concentric spheres from their point of origin. A laser beam would be a tight, narrow shaft of

119   "[Deleted]")

light, so that no one could tell what message it carried unless standing right in the path of the beam. Such a communications system could increase the security of military messages a millionfold.*

Nevertheless, the U.S. government seems sold on laser "guns," and presently indicates that it will press on with development of offensive and defensive laser weapons. Late in 1980, the Senate Commerce, Science and Transportation Committee issued a report that urged the U.S. to push ahead with a *"Star Wars* defense plan," as the Boston *Herald American* described it, including a laser anti-missile system and a greatly increased budget for laser weapons research. The report suggested that laser technology could transform the international balance of power and inaugurate a "post-nuclear" era, with all such a change might entail. Anti-missile and ASAT lasers would, if successful, mean a big change in politics among the superpowers. A nuclear arsenal might no longer serve to deter aggression—and, on the other hand, nuclear weapons might no longer be useful as aggressive weapons, either. Advances being made in American and Soviet laser labs today may soon reshape our world as thoroughly as progress in physics laboratories did a generation ago.

* This kind of communications system may also revolutionize civilian telephone systems by eliminating the need for thick electrical cables.

NINE

"Anything but the Truth"

EMPEROR Marcus Aurelius once wrote, "I cannot comprehend how anyone can desire anything but the truth." Desiring the truth is easy. Tracking it down can be much more difficult—especially in the controversy over particle-beam weapons (PBWs), a fantastic new array of "guns" being considered for the United States arsenal.

PBWs would be something like the disintegrator rays of science fiction. Instead of shells or bullets, they would use as ammunition the tiniest objects known to scientists: subatomic particles.

In classical atomic theory, all matter is made up of atoms. Democritus, the founder of that theory, said there were ultimately but two realities in the universe—atoms and empty space. Yet Democritus might have been startled to learn that his atoms are mostly empty space themselves. Only a tiny fraction of an atom's volume is made up of particles; and if one squeezed all the vacant space out of the earth's atoms, the whole planet could be reduced to the size of a golf ball.

The particles that comprise atoms are called electrons, protons, and neutrons.* The electron, the negatively charged particle, is so small and insubstantial that there is some debate about whether it can rightfully be called a particle at all. It has virtually no mass and is about 10^{-12} centimeters wide. If you took one electron for every human being on earth and laid all the electrons out in line, they would not reach across a single hair of your head.

Electrons exist in the fuzzy borderland between physical objects and pure mathematics. It is becoming fashionable to think of them less as particles than as "standing

* Subatomic physics, which deals with what lurks inside the atom, has undergone a revolution in the last twenty years. Our old three-particle view of the atom's structure has been modified almost weekly, as physicists discovered new particles with the help of high-energy "atom smashers." Now there is a long list of particles with bizarre names and sometimes Greek-letter prefixes as well—"pi-mu mesons," for example. These particles can be so hard to recall that Enrico Fermi, who helped to build the first functioning nuclear reactor at the University of Chicago, once said in a moment of exasperation: "If I could remember all these particle names, I would have been a biologist!" For the purposes of this discussion, however, the three-particle model of the atom will suffice.

waves"—that is, stationary vibrations—in "shells" of energy that make up the outer layers of the atom.

Still, electrons behave in a particulate manner. They may be added to and subtracted from atoms like change from a cash register. They may be fired like bullets, scattered like grass seed, bounced off targets like handballs off a court wall. Their opposite numbers in the atom, so to speak, are the protons.

Over 1,800 times more massive than electrons, protons each have the same amount of electrical charge as an electron, but a positive charge rather than a negative one. Protons nestle in the core, or nucleus, of the atom, while electrons whirl about the nucleus like bees about a hive. (Actually, physicists prefer to think of the electrons in terms of a "probability cloud"—each "droplet" in the cloud representing the chance of finding an electron there at any given moment. But the bee-and-hive model is simpler and easier to understand, and is still widely used.)

The number of protons in a nucleus determines which element the atom will be. One proton—hydrogen. Two protons—helium. Three protons—lithium. Seventy-nine protons—gold. Ninety-two protons—uranium. Ninety-four protons—plutonium.

In an electrically neutral atom, there are exactly as many electrons as protons, so their charges balance out. But these particles are not alone in the atom. In every element save simple hydrogen, they have the company of a third subatomic particle: the neutron.

Neutrons are just as their name implies: neutral. A neutron has no charge. It is the Switzerland of subatomic particles. A neutron has a shade more mass than a proton, and can be added to a nucleus to change the atom's weight and "isotope." The isotope number indicates how many protons and neutrons together exist in the nucleus.

An element remains the same when its isotope changes, because the number of protons stays stable. Uranium, for instance, always has 92 protons. A variation of a few neutrons, however, can have drastic effects, on humans as well as on the atom. Uranium-238 is stable and harmless. Its cousin uranium-235 is unstable and highly explosive. A few chunks of uranium-235 wiped out Hiroshima in 1945. Three neutrons can make that much difference.

Their lack of charge gives neutrons useful properties. For one thing, they are unaffected by electrical fields. Shoot a charged particle—a proton or electron—into such a field, and it will be deflected. Neutrons, however, are oblivious to the field's attraction. They sail straight through, their course unchanged. How this trait makes neutral particles—in theory—a handy weapon, we shall see shortly.

Physicists use a "gun," called an accelerator, to fire subatomic particles. By sending a charged particle through a rapidly alternating series of positive and negative fields, physicists can force the particle to be "kicked" and "tugged" (depending on the field's polarity) up to velocities close to that of light—about 300,000 kilometers per second. A stream of such particles may hit a target with devastating effect.

Particle-beam weapons first made headlines in the United States in 1977. An article in *Aviation Week* revealed evidence that the Soviets had built a giant particle-beam projector in the southern U.S.S.R. An editorial in the same issue voiced concern that the Russians might soon have a functioning PBW capable of "checkmating"

U.S. defenses by wiping out the whole American arsenal of ballistic missiles.

Released about the same time as the motion picture *Star Wars*, news stories about Russian "death rays" and "zap guns" were printed along with illustrations from the film, and fed on the movie's publicity. A typical headline appeared in the *New Republic:* "STAR WARS FOR REAL?"

Soon wild reports were flitting like bats through the media. A giant Soviet PBW was supposedly being readied for launch into orbit, while the U.S. tried desperately to catch up with Russia's enormous lead in particle-beam technology. Before we could match them particle for particle, the Soviets were going to turn their infernal space gun on our satellites and make them disintegrate in a shower of subatomic "bullets."

Failing that, the Reds were simply going to aim their orbiting PBWs down into the atmosphere over U.S. cities and amortize millions of innocent civilians in a shower of secondary radiation. The space shuttle was going to get it, too. It would fly up and run straight into a lethal shower of protons, electrons, or neutrons.

The Defense Department was understandably concerned about this publicity, because it left the public with the impression that little or nothing was being done to counter a serious strategic threat from the U.S.S.R. So on May 3, 1977, DOD attempted to reassure the public in an official statement to the media:

> Senior officials of the Defense Department do not believe that the Soviet Union has achieved a breakthrough in research which could soon provide a directed-energy beam weapon capable of neutralizing ballistic missile weapons. Based on all information now available to the

U.S. intelligence community, this possibility is considered remote.

Not completely remote, however, for a DOD fact sheet released that same year indicated the Soviet beam research effort was indeed something to be reckoned with:

> The Soviet effort on charged particle beams is judged to be larger than ours, particularly in the area of accelerators for future fusion applications. However, their state of the art is approximately comparable to ours. In some areas they lead, and in others the U.S. leads.

Still, the official view from Washington was that the Russian PBW threat was dubious. Defense Secretary Harold Brown said in 1977 that he was "convinced that we and they can't expect to have such a weapon system in the foreseeable future." After all, the United States had been working on PBW technology for more than a decade without any advances significant enough to put a PBW "cannon" into space. Brown admitted, however, "I can't predict how things will be 20 or 30 years from now."

It is difficult to tell how things are even today, where the state of PBW research is concerned. Nearly all the militarily important research is classified, and the little information that seeps out to the news media may be colored to present a misleading picture. Nonetheless, it is widely assumed that the U.S. is making steady progress in beam propagation, and the next few years may see a PBW capable of knocking missiles and satellites out of the sky.

Behind the improbable code name of Chair Heritage hides the Navy's PBW project. It is a highly sophisticated defense system for large ships like aircraft carriers, which stand out like islands in the ocean and make splendid targets for enemy missiles and aircraft.

A modern cruise missile, unlike its primitive ancestor the V-1, is a damnably hard weapon to intercept. It can fly so low as to avoid radar detection. It can maneuver in order to avoid interception. It is small, and thus hard to target. And a cruise missile is inexpensive as weapons systems go these days, so that perhaps a hundred cruise missiles, each equipped with a nuclear warhead and capable of sinking a ship, can be launched for the cost of a single jet fighter-bomber.

Defending a carrier from that kind of attack is a nearly impossible job if one relies on conventional anti-aircraft and anti-missile weapons. First the attacking missiles must be targeted. Then other missiles must be launched to intercept them, and the crew on board the ship must wait and pray that their tracking is good enough to destroy the enemy's weapons. If not, then the ship and the several thousand persons on it will vanish in a nuclear fireball.

One cruise missile could accomplish this. If the enemy launches a large salvo of them, the probability is virtually 1.000 that at least one missile will penetrate the ship's defenses and explode on target.

Suppose, however, that the vessel were guarded with anti-missile weapons that traveled at the speed of light and could destroy a target almost instantaneously. That, in a sentence, is the goal of Chair Heritage. A ship-based PBW could knock down cruise missiles—or aircraft—as fast as they were launched.

As soon as radar spotted the incoming blips, a beam of particles would rip out of a turret on the ship's upperworks at Einsteinian velocities, and fry the first missile

0.00006 second later. In another split second, the next missile would go up in flame. Then the third, the fourth, and so on, until the sky was clear of hostile targets.

There would be plenty of time to zap them all, because the particle beam would travel at almost 300,000 kilometers per second. Nothing could outrun or outmaneuver it. Once in the sights, a target would be as good as gone.

The Navy is not alone in its fascination with PBWs. The Army has a PBW research effort under way as well. Originally it was named Sipapu, a native-American word meaning "sacred fire." Out of deference to the sensibilities of the original Americans, the name has been changed to White Horse.

This project has been consolidated, along with all other military PBW work in the United States, under the umbrella of DARPA. Its purpose, however, remains the same: to develop a "neutral-particle-beam accelerator" that could strike down missiles or satellites thousands of kilometers away.

Using sometimes scanty Russian data on PBW research, American scientists are reportedly working on a device called the "RFQ accelerator." "RFQ" stands for radio frequency. Though the details of the device and its workings are extremely complicated, basically the accelerator's job is to push particles toward light speed more efficiently than previous accelerators have.

In effect, White Horse is preparing a new and better "charge" for the PBW "gun." It is designed to make particle-beam weapons simpler and more durable, and thus better suited for placement in space.

What makes White Horse a good candidate for this job is the lack of charge on its ammunition. A charged-particle beam, as noted earlier, could be swayed off target by an electrical or magnetic field. And since the earth is a gigantic natural magnet, it fills the surrounding space with long curved lines of magnetism that could lead a charged-particle beam astray, just as contrary winds at sea may blow a ship off course.

With neutral particles, that problem never arises. A magnetic field makes no difference to them. They rip through it like bullets through a fog.

Yet no one has written off charged-particle beams as unfeasible. The Soviet literature on physics indicates that the U.S.S.R. is looking closely at electron beams as weapons and may be nearing development of an effective charged-particle weapon.

Reports reaching the West have it that the Soviets are testing electron beams on targets made of aluminum alloy and explosives; that is, simulated conventional weapons. If the Russians really are conducting such experiments, then they probably mean to use PBWs on the battlefield.

Also, Western analysts think it significant that physicists specializing in charged-particle-beam weapons have recently been admitted to the Soviet Academy of Sciences. It is hard to imagine the Russians giving those scientists such a high honor unless their work had a serious military application.

Consequently, electron beams are the object of intense interest in the United States. As yet, the beams are too weak to have any immediate use in warfare, but there are plans to escalate them to what DOD calls "weapons-level intensities," at which the beam will become a giant artificial lightning bolt.

PBWs have their critics in the scientific community,

however, and a number of influential voices in American science have strongly disputed the claims made for particle-beam weapons. In 1979, four faculty members of the Massachusetts Institute of Technology produced a report for the MIT Program in Science and Technology for International Security (PSTIS) that debunked the military potential of PBWs. The report concluded that:

Tremendous amounts of energy would be required to generate a particle beam capable of doing its target any harm.

The beam would have to be aimed with supreme precision, and it is not certain whether such accuracy is possible.

PBW attacks might be defeated by cheap and simple countermeasures.

The MIT study points out that there is an energy threshold, one might call it, below which a PBW will be useless, because it will not hit the target hard enough to do any damage. That threshold, according to the report, is about 1,000 joules per square centimeter.*

A stick of dynamite releases some 300,000 joules when it explodes. The PBW will therefore have to concentrate that much energy on an area about equal to one side of a business envelope if the particle beam is to be effective as a weapon.

* A joule is a metric unit of work, and is defined as the work done when something is moved one meter, in the direction of the force, by the application of one newton of force—a newton being the amount of energy required to accelerate one kilogram of something one meter per second—per second.

The particle beam would confound an ICBM's guidance system, or touch off its fuel tanks or its warhead. It can do this, however, only if those 1,000 joules per square centimeter are delivered *inside* the missile. What counts is not how much energy strikes the surface of the target, but how much penetrates the skin and reaches the vulnerable "entrails."

To put an adequate dose of deadly force inside a missile, the PBW will have to deliver much more than that to the exterior, because even a few millimeters of metal will intercept most of the energy of the particle beam. And, assuming the aluminum hide of a missile is five millimeters thick (about the width of an ordinary pencil), then the MIT team's calculations require the PBW to land some *30,000 joules per square centimeter* on the missile's skin. That's the energy of our exploding dynamite stick, focused on an area the size of a sugar packet.

Generating a beam that powerful would require tons of chemical fuel, or a sizable nuclear explosion. For purposes of argument, however, let us suppose that such a beam can be generated. What happens to the beam on the way to the target?

If the beam consists of charged particles, they may be deflected by the earth's magnetic field or by other, more localized magnetic or electrical fields. Even a small nudge (a tenth of a second of arc, for instance) might carry the beam away from its quarry, especially if the target is less than a meter or so wide. (A cruise missile is about the diameter of a telephone pole.)

For the beam to be effective, it must score a *direct hit*. An interceptor missile, by contrast, need not hit the bull's-eye to achieve a kill. If an explosive warhead goes off in the near vicinity of a cruise missile or an ICBM, the blast will probably be sufficient either to force the missile off course or to wreck it completely.

Another problem with particle beams is similar to thermal blooming in lasers. The atmosphere gets in the way of the beam.

To travel through the air, a particle beam would have to ionize the atmosphere around it, thus creating a "tunnel" for the beam's passage. Ionizing the air would siphon energy away from the beam, however, and for that reason the MIT study suggests that PBWs may pose a threat only at distances of a kilometer or less.

But for the moment, let us discount all atmosphere-related problems. Assume there is no trouble in shooting the beam through the air. There are *still* defenses against a particle beam.

If targets are tracked optically, it would be simple and inexpensive to hide them behind a smoke screen. The contents of a container of magnesium oxide no bigger than a coffee can, when released into the air, will hide an aircraft carrier from view. The same amount could surely conceal a couple of incoming missiles. Clouds would also provide a natural smoke screen.

If the weather is clear, the attacker has another option. Inasmuch as the particle beam requires a corridor of ionized air to let it pass, a well-placed explosion that interferes with that corridor will cause the beam to fail.

So many obstacles to PBW warfare are atmosphere-related that outer space would seem to be the best place for them. There no atmosphere exists.

In theory, a spaceborne PBW should be able to pick off enemy satellites with ease. Here, however, fuel is a limiting factor.

To generate particle beams capable of damaging an enemy, a PBW must carry a tremendous source of power. One salvo could consume tons of fuel. Even nuclear bombs—assuming their energy could be harnessed to run a PBW platform in orbit—would not satisfy the machine's appetite for long. Presumably, the station's tanks could be refueled at intervals by spacecraft sent up from earth, but such missions would be prohibitively expensive. For long-term use, it is hard to imagine an orbiting particle-beam weapon equipped with any feasible energy source except controlled fusion; and that kind of power seems decades away.*

As for the threat of orbiting PBWs turning their beams on the earth below and soaking soldiers or civilians in the open with secondary radiation, this kind of attack is highly improbable.

According to calculations from a recent issue of *New Scientist,* even if one assumes a PBW in space could deliver some 18,000 joules per square meter to the ground, over an area 20 kilometers or so in diameter (which would be the minimum performance for the weapon to be useful in war), and could generate enough secondary radiation to kill a soldier in an hour (any longer than that, and the troops could save themselves merely by driving out of range), the particle-beam weapon would

* A fusion reactor would squeeze hydrogen atoms together, under tremendous heat and pressure, to form helium—like the process that keeps the sun shining. Controlled fusion would make available a titanic new source of energy, because hydrogen is the most abundant element in the universe. Fusion would also be relatively pollution-free compared with conventional fossil fuels and nuclear fission, all of which generate harmful wastes. The only "waste" from a fusion facility would be safe helium.

need approximately *one million* power units fueled with nuclear bombs.

This is clearly a cost-inefficient method—to use the Pentagon's own jargon—of killing the enemy. And while a groundward-pointing PBW satellite an order of magnitude less powerful than the one just described might be built and aimed at armies in the field, the radiation dosage would be so slight that weeks would pass before the soldiers began to feel sick. By that time the war might be over.

If particle-beam generators in space do prove to be effective weapons, a prototype PBW platform might be built in orbit. The station would have to be large, however, given the size of the accelerator and the energy sources needed to power it; and anything as big as a spaceborne PBW would surely be detected by the Soviets.

Moreover, any significant step toward installing a PBW in space would be likely to spark Russian retaliation in the form of an ASAT strike. The Russians have demonstrated their ability to knock down comparatively small targets; neutralizing a giant construct like a particle-beam generator—which would probably have to be at least as big as the Skylab space station—would be as easy as hitting a barn with a baseball. And if the Soviets started assembling a PBW in orbit, the United States would have reason to respond in the same way.

Nonetheless, the PBW race continues. In the summer of 1980, a new series of reports surfaced in the U.S. press about Soviet progress in PBW work. *Aviation Week* published details of a charged-particle-beam weapon that had reportedly been built at Saryshagan, in Kazakhstan.

Code-named Tora, the PBW at Saryshagan is said to be

powered by about a dozen "magneto-explosive generators," cannon-shaped devices which transform the energy of chemical explosions into electromagnetism. The generators feed their power to a switching room, which in turn funnels it to a particle accelerator. The electron beam created in the accelerator shoots out an "ejection nozzle" toward the target, and is aimed by magnets that train the beam in the desired direction.

This is only one interpretation of the hardware at Saryshagan. It may also be seen as a superlaser instead of a particle-beam generator. But, in any case, the Saryshagan installation has the United States worried, and has added new impetus to the American PBW program.

If Russia has—or soon will have—a working PBW, the U.S. would be well advised to build one too. And even if the Soviet beam weapon turns out to be only a phantom, then the work and cash expended on U.S. particle-beam technology may not be wasted completely. The technology used to build a PBW may have other useful applications, particularly in fusion energy research.

At the same time, however, it is hard to escape the feeling that the PBW "gap" trumpeted in the press is, at bottom, little more than an excuse to spend military dollars.

As *New Scientist* suggested in an article on beam weapon research, the publicity given PBWs "seems to be part of the big lie by which the military generates funds to provide 'defence' against a non-existent threat." Indeed, the military has done just that on numerous occasions in recent history. Since the end of World War II, Americans have been warned about the dangers of a missile gap, a tank gap, a navy gap, an anti-submarine warfare gap, a chemical warfare gap, a biological warfare gap, a bomber gap, an infantry gap, an electronic warfare gap. . . . The list would fill an entire page. Many of these alleged gaps have turned out to be illusory—or, if they existed at all, to be grossly in America's favor. The missile gap so feared during the Kennedy years is a good case in point. During the early 1960s, when the administration and the media were fretting publicly over a perceived disparity between Soviet missile might and ours, the Russians were in fact nowhere near as strong in ICBM "throw weight" as the U.S. was.

Perhaps the PBW gap will go down in history as just one more illusion—or perhaps it is as real and perilous as the German rearmament just prior to World War II. The truth probably lies somewhere between these two extremes, and only time will tell where.

TEN

"The Kiss of Death"

AT FIRST, television screens all across America showed only the blue sky. Then a fuzzy lozenge appeared, swinging back and forth across the field of view as the cameramen on the ground tried to keep it in sight. Eventually viewers could make out the distinctive outline of the NASA space shuttle Columbia, returning from its first mission into space.

A small army of technicians, NASA officials, visiting dignitaries, and journalists waited to witness Columbia's

137)

landing on the flat, hot bed of Rogers Dry Lake in the Mojave Desert. Every eye and every lens was trained on the descending spaceship as it banked and glided into its final approach.

Chase planes in attendance, Columbia touched down gently on the desert and rolled to a stop, trailing a huge cloud of dust.

"Beautiful," said mission control. "Welcome home, Columbia!"

Governed by computer, the launch had gone flawlessly. Columbia rose from its pad at Cape Canaveral, Florida, on the morning of April 12, 1981, supported by two pillars of brilliant white flame from its two solid-fuel booster rockets. Bolted to the shuttle's underbelly, between the solid-fuel boosters, was the external fuel tank, fifteen stories tall and holding 350 tons of fuel for the shuttle's main engines. At first glance, this awkward assemblage of tank, rockets, and spaceship looked as if it would never fly. But it flew superbly; and Columbia's success sent America into a near-delirium of joy and self-congratulation.

"SUCCESS," said the front-page banner headline in the Boston *Herald American*. "AMERICA'S GEM COMES HOME," crowed the Boston *Globe*. Similarly, the Washington *Post* pronounced Columbia the "GEM OF THE SKY," while a New York *Post* headline said "AMERICA LOVES YA!" above a photo of astronauts John Young and Robert Crippen.

Like John Glenn's orbital journey almost two decades earlier, the shuttle's first mission became the excuse for all kinds of commentary. In editorials and columns, NASA's spaceship was used to illustrate everything from our religious heritage (one newspaper, in a moment of hyperbole, called the shuttle a "flying cathedral") to the need for upgrading United States public schools. How, one columnist demanded, can we keep shuttles flying if our children, the astronauts of the future, are never taught to read, write, and calculate?

Nothing about the shuttle was too trivial for media coverage. The Washington *Post* reported that the shuttle's captain had been granted legal authority to perform arrests in space.

Almost buried in the avalanche of amusing and inspiring news about the shuttle were a few serious stories, scattered here and there through the media, about the shadowy side of the shuttle project—its military role. Although NASA has tried to present the shuttle as a basically peaceful project with some military uses, the fact is that DOD has all but taken complete control of the spaceship. One top Air Force official was quoted in 1981 as saying that NASA was only "a minor user" of the shuttle, and "not the driver."

The Pentagon had had a keen interest in the shuttle from the moment it was conceived. Around 1970, Project Apollo was winding down. The United States had met President Kennedy's goal of landing a man on the moon before the end of the sixties. NASA knew that when Apollo ended, some new job would have to be found for the space agency. And given the cost-cutting, hardheaded mood of Congress and the Nixon administration toward space travel, NASA knew it would have to devise some duties for itself that were more practical than Apollo's costly joyrides to the moon.

So NASA turned to the old idea of a manned, winged

space glider for its meal ticket. It was, after all, the kind of spaceship NASA had originally sought. Only the deadline pressure set by Kennedy had forced the agency to abandon that plan and concentrate instead on ballistic capsules.

Unlike the capsules, the space glider could be used repeatedly. There was no longer any need to build costly, wasteful one-shot spaceships. Theoretically, the glider would be far more cost-effective.

Also, recovery would be simpler. The glider would slip down through the atmosphere under a pilot's control and land as a conventional airplane would. The Navy would not have to deploy a whole task force to recover the spacecraft from the sea. This flexibility would increase the spaceship's safety, too. If one landing area was inaccessible for some reason, the pilot could simply pick another.

Finally, there were all kinds of useful jobs the manned space glider could perform in orbit, from pollution surveys to deploying satellites. Rather than being launched individually, each one atop a separate and costly rocket, satellites might be sent up, five or ten at a time, in the spacecraft's cargo bay, and released into orbit.

In short, NASA had in mind something like the Air Force's X-20. So the military began to take an extravagant interest in NASA's space shuttle.

Just as with the Gemini project years earlier, NASA agreed to cooperate with the Air Force on the shuttle. One of the first requests the Pentagon made was for a change in the cargo bay. NASA had envisioned a shuttle with enough space on board for perhaps a couple of tons of equipment. For scientific purposes, that would have been sufficient. DOD, however, extended the cargo bay to a length of 18 meters, so that the shuttle could haul roughly 20 tons of payload. That is the equivalent of ten private automobiles, or a fully loaded moving van.

There was no way to justify such an enormous cargo capacity on scientific grounds. After Apollo and Skylab, only the Defense Department was likely to do anything in space that would call for a 20-ton delivery. Was the Pentagon thinking of reviving MOL?

NASA's alliance with DOD disturbed some members of Congress. One unnamed figure on the Senate Committee on Aeronautical and Space Sciences was quoted in *Science* as saying the Pentagon's close relationship with the civilian space agency might prove to be "the kiss of death" for NASA. But NASA remained the Pentagon's partner.

Under DOD's direction, the shuttle has grown from a relatively small research vehicle to the biggest and most sophisticated spacecraft ever flown, causing all manner of headaches for NASA. First there was the matter of the tiles.

The shuttle's hull is covered with thousands of individual ceramic tiles for insulation against the terrific heat of re-entry, when the air just ahead of the hurtling spaceship glows hotter than the sun's surface. The tiles are a marvel of engineering. They convert kinetic energy into heat and dissipate the heat so efficiently that one can pick up a red-hot tile in one's bare hand without being burned.

Unfortunately, the tiles are also brittle and easily damaged. Their fragility became apparent, to NASA's dismay, when Columbia was flown from California to Florida on the back of a 747. Strips of tape on the shuttle's exterior worked loose and slapped at the hull in the airplane's slipstream, breaking and dislodging many of the tiles. Columbia arrived in Florida in sad condition. Huge ugly brown gaps showed in the white insulation, where clusters of tiles had dropped off.

The tiles were only one problem. A balky engine on the orbiter delayed the maiden flight repeatedly. Solid fuel in the boosters sagged out of shape. As troubles mounted, the initial flight had to be postponed from 1978 to 1979, then 1980, and finally 1981.

Meanwhile these delays angered Congress, which had to appropriate NASA's annual budget and expected to see some results for its money. Instead, NASA kept coming to Capitol Hill, hat in hand, asking for more cash to bail out its troubled project. Long before the first craft left the launch pad, the shuttle was running 20 percent over its projected budget.* Moreover, the space agency was unsure when the first flight *would* take place. At the rate problems were developing, the maiden flight might not occur until well into the 1980s.

In 1979 alone, Congress had to give NASA $185 million in additional funds to keep the shuttle alive, and the project was expected to cost $1 billion more than anticipated (in constant 1971 dollars) to get the spaceship off the ground. The legislators began grumbling, and wondering aloud if the shuttle was worth it. What the shuttle needed was Presidential backing; and while President Jimmy Carter had expressed his interest in the shuttle on several occasions—during one televised interview, he had said he hoped to ride in it someday, if his age allowed—he had never given space projects his fervent support.

His attitude changed quickly when the Pentagon

* A few years ago, that difference would have been called a "cost overrun." Now overruns are so common that they are called "cost excesses."

briefed him, in November of 1979, about the military use of the spaceship. At once Carter spoke up on the shuttle's behalf, and called on Congress to approve NASA's request for more money. Congress complied. For fiscal 1980, the shuttle received an extra $220 million—almost 20 percent increase over the previous bail-out check.

No decision of any recent Presidential administration more clearly illustrates the federal government's policy of pushing military space projects at the expense of civilian ones. At the time those extra hundreds of millions were being showered on the shuttle to ensure that DOD would have it, non-military space projects such as the Galileo probe to Jupiter were being postponed or dropped entirely because the government said the money for them simply could not be found. The dwindling number of scientific job allotted to the shuttle, in the face of DOD's demands and the project's budget pinch, prompted Senator Adlai E. Stevenson III to ask at a 1979 Senate committee meeting: "Are we in a situation where this country finally achieves an operational space transportation system, only to have lost the capability to use it for science?"

In these days of restricted funding for non-military space exploration, NASA has had to face the unpleasant possibility of its own extinction. Indeed, the Pentagon's interest in the shuttle may be the only reason NASA exists today at all. The shuttle is NASA's reason to be. And, as noted earlier, the shuttle's military potential is apparently all that saved the project from cancellation. So NASA has been forced to take on DOD as a senior partner, if not boss, and in the process has become little more than a Pentagon fiefdom.

In the meantime, some of the claims used to sell Congress originally on the shuttle as a peaceful, productive "workhorse" of space have come to sound hollow. In a 1980 article on the shuttle, the *Washington Monthly* pointed out that there is no need to retrieve communications satellites when they malfunction; that such a mission would be folly for two reasons. First, the benefit would not be worth the risk of the assignment. It made no sense for NASA to expose its highly expensive spaceship, and its crew, to the hazards of spaceflight for the relatively trivial task of plucking a defunct civilian satellite out of orbit for maintenance. Even if NASA was willing to handle such a job, the businessmen who own those satellites would be foolish to want them returned to earth in the first place; it would be more cost-effective to send up a new satellite, because the old one would probably be obsolete by the time its lifetime ended. Finally, most of the satellites that anyone might wish to have retrieved are in geosynchronous orbit more than 30,000 kilometers out in space—and the shuttle is designed to fly at an altitude of only a couple of hundred kilometers.

That same article pointed out that NASA's estimates of cost for a typical shuttle mission are highly optimistic.

Since 1975, NASA had contracted for shuttle missions at a rate of $22.4 million per flight. To launch a single satellite on a conventional NASA rocket such as the Delta, the bill will come to somewhere between $25 million and $35 million.

That $22.4 million figure was based, however, on NASA's assumption that each shuttle would fly into space on the average of once a week, for a total of some fifty space missions per year. The private sector has nothing like that kind of demand for launches. Perhaps two or three civilian comsats are launched in a typical year, though that figure may rise slightly in the 1980s as more and more countries build their own satellite broadcast systems. If NASA is to break even, then, the price of an individual launch will have to multiply greatly to about that of the one-shot rockets.

The private sector appears to have little incentive to use the shuttle. By far the biggest single share of shuttle missions will go to DOD.

It is difficult to say exactly what kind of work the Pentagon has planned for the shuttle. Details are shrouded by official secrecy. Some areas of DOD interest have been reported, however, in the aerospace trade journals. For examples:

Satellite deployment. For satellites that will orbit higher than the shuttle flies, plans call for the shuttle to carry an unmanned rocket called the Interim Upper Stage (IUS), which will lift the payloads up to geosynchronous orbit. Some of these satellites will be active from the moment they are deployed. Others, rumor has it, will be stationed deep in space and kept in reserve for use in case satellites in lower orbits are attacked. These high-flying "dark" or "silent" satellites, as they are called, would be beyond the reach of ASAT missiles, but might be targets for lasers or particle beams.

Heavy payloads. The Pentagon would like to see the shuttle lift still more than it can now. That is probably one of the reasons NASA is looking for ways to increase the shuttle's cargo capabilities. The space agency is investigating a scheme to add a liquid-fuel "thrust augmentation" system—an extra booster stage, attached to the external fuel tank—to the shuttle, bringing its potential payload weight to almost 50 tons. Hundred-ton cargoes are under consideration for future shuttles. If DOD intended to build manned space stations or PBWs in orbit, such a heavy lift capability would come in handy.

Space tankers. One restriction on past manned space missions has been fuel. There were no service stations in the sky, and the spaceships had to carry all the fuel they might need. Now the shuttle may be able to remedy this situation. The Johnson Space Center is reportedly looking for ways to convert the shuttle into a spacegoing fuel tanker, presumably to supply future manned space stations—or, conceivably, to top off the tanks of orbiting laser weapons and PBW platforms. This refueling capability would make spaceborne particle-beam weapons a more attractive proposition. NASA is also considering building refineries in space, where water hauled up from earth would be broken down by electrolysis into hydrogen and oxygen.

Large structure studies. In 1980, *Aviation Week* reported that NASA is cooperating with the Pentagon on a demonstration project for building big assemblies in space. How big is anyone's guess. NASA has spoken in recent years of building space stations the size of resort hotels for peaceful purposes, and the armed forces may have dreams exceeding even that scale.

Spacelab. NASA says there will be twenty flights a year carrying the European Space Agency's (ESA) Spacelab, a manned laboratory that will fit inside the cargo bay and will be pressurized to give astronauts working inside it a comfortable, shirt-sleeve environment. The lab will be connected to the cabin of the space shuttle by a small access tunnel. Thus far, NASA has been vague about the jobs Spacelab will perform. In *Space Shuttle,* an official NASA booklet, the text says only that "many types of scientific, technological, medical, and applications investigations can be accomplished with this flight hardware." Might those technological and applications investigations include work for the military? Spacelab does sound very much like the Air Force's old MOL, the military space station scrapped because of the expense of Vietnam.

What are the chances that the shuttle will serve as an ASAT weapon? Very small. The shuttle would have some drawbacks as an anti-satellite system.

For one thing, it is big. A spacecraft the size of a medium-sized jetliner would make a splendid target for any Soviet ASAT weapons. The shuttle also flies low, making itself even easier pickings for enemy weapons. And if it were ever sent up to intercept Soviet satellites, it would be easy for the Russians to mount countermeasures.

As Dr. Richard Garwin of Harvard states in a 1981 article for the *Bulletin of the Atomic Scientists,* a Russian satellite could be booby-trapped to foil any tampering by an American shuttle. The Soviets could merely install an explosive charge on their satellite, hook the bomb to an acceleration sensor, and set the detonator to go off if the satellite is nudged by anything except its own rocket thrust. The shuttle flies up, reaches out with its long manipulator arm to seize the Russian satellite—and the bomb explodes, wrecking both satellite and shuttle.

Of course, the shuttle might stand off at a respectable

distance and bombard its target with missiles. Yet such a job could be accomplished much more quickly, cheaply, and safely by unmanned spacecraft like the missile described in the opening pages of this book.

The shuttle, in short, seems unlikely to see duty as a satellite killer. Nonetheless, the Soviets have accused the U.S. of building it for just that purpose.

In the spring of 1979, the Soviet Union asked the United States to halt work on its space shuttle as a step toward the demilitarization of outer space. Predictably, the Americans refused. The *New York Times* quoted an unnamed U.S. official close to Soviet-American arms limitation talks as calling the Russian request "unacceptable" and adding that the United States "would never agree to terminate the shuttle or even to slow it down."

As *Aviation Week* pointed out, the Soviets' adamant stance on the issue of the shuttle's ASAT potential may be just a bargaining tactic used in the hope of forcing other concessions out of the U.S. at the negotiating table. Still, the shuttle's military capabilities appear to have the Kremlin badly worried. After Columbia's first flight, Soviet President Leonid Brezhnev dropped heavy hints to that effect at a celebration honoring two cosmonauts, a Soviet and a Mongolian, recently returned from orbit. Brezhnev said in his speech:

> "I should like to stress that the Soviet Union has been and remains a convinced supporter of the development of businesslike international cooperation in outer space. May the shoreless cosmic ocean be pure and free of weapons of any kind. We stand for joint efforts to reach a great and humanitarian aim: to preclude the militarization of outer space."

Brezhnev let the timing of his remarks convey Russia's concern over the U.S. shuttle. General Vladimir Shalatov, chief of the U.S.S.R's cosmonaut training program, voiced his concern about the American shuttle's military role more bluntly, at a press conference held several days before Brezhnev's address.

Shalatov said the Soviets were "alarmed that there is less talk of the economic benefits of the shuttle, but more is being heard about military uses." He may have been thinking of what Defense Secretary Harold Brown had told the Senate Committee on Commerce, Science and Transportation on February 7, 1980. "Over the next five years, our dependence on the shuttle to support our space systems will become critical," Brown said. "Space systems are increasingly more important for support of our military forces in areas such as communications, navigation, early warning, surveillance and weather forecasting." Brown added that the shuttle was expected to provide "the benefits of reduced launch costs, increased reliability, increased weight and volume of our payloads and, perhaps most important of all, increased flexibility." In other words, DOD expects the shuttle will allow the Pentagon to send up more and bigger military hardware for less money per pound than ever before, and that it will make practical many military operations that were previously difficult or impossible.

Representative Ron Paul spoke more plainly in a short speech on the House floor on June 21, 1979:

"Mr. Speaker, the Soviets would like very much to see an end to the space shuttle program, which is a good reason to expand it.

"The space shuttle has tremendous defense implications, which is why the Communists want to get rid of it.

"NASA's future, I believe, must lie in defense-related capabilities. The most important thing we can do to achieve this is to enlarge the shuttle program.

"And . . . we ought to consider arming the shuttle with defensive weapons.

"The space shuttle is far too valuable to our nation's security to risk its being destroyed by hostile enemy action in space."

Paul's remark about NASA's future was an ominous sign for the space administration, for it indicated that Congress—in the past a strong backer of the civilian space effort—is thinking of letting DOD "eliminate the middleman," so to speak, by simply taking over NASA. At present NASA is involved so deeply with Defense Department business that a 1978 Library of Congress study seriously discussed a plan to hand over NASA's launching operations to the Pentagon. And an Army War College report made public recently under the Freedom of Information Act refers to America's separate civilian and military space programs as "expensive . . . luxuries" that the nation cannot afford much longer. That report called for an "integrated civil/military space technology program," thus proposing that the Defense Department swallow up the civilian space effort.

The Pentagon is now the virtual master of NASA in fact, if not in name. DOD controls the shuttle, and the shuttle is NASA's main reason for existence now that the glory days of Project Apollo are gone. The situation is different from what it was in the Gemini years, when the Air Force was trying to get a foot in the door at NASA. Now the door is wide open, and the Pentagon is making NASA's house its own. The result is that we face the likely extinction of America's non-military space program within the current decade—a chilling object lesson in what can happen when the lamb lies down with the lion.

ELEVEN

"Dynasoarski"

THERE IS an interesting symmetry to the United States and Soviet manned space programs. We have had three manned space capsules: Mercury, Gemini, and Apollo. They have had three manned space capsules: Vostok, Voskhod, and Soyuz. We have had a manned space station: Skylab. They have had a manned space station: Salyut. And so, when the U.S. space shuttle was being developed, NASA and DOD naturally wondered if the Soviets would match that effort with their own shuttle. Rumors about Soviet shuttles had been floating about the

U.S. intelligence community for years, starting with the report of Russia's T-4A back during the early 1960s. But there was no solid evidence to back up these stories. If the Soviets were working on a shuttle, they were unlikely to spread the word until they had their spaceship well on the way to routine service.

The Soviet Union surely had good reason to want such a vehicle. A reusable shuttle would help the Russian space effort in the same ways NASA's orbiter was expected to help ours: by lowering costs and increasing "flexibility," as Defense Secretary Brown put it.

Moreover, the Soviets needed some way to make manned missions safer. On launch and in orbit, Russian spacecraft were no more dangerous than their American counterparts. But landing was more risky for the Russians than for U.S. spacemen.

American re-entry vehicles could splash down in the sea, and the water would cushion their landing. Then the capsule would float until the Navy could recover it and take the astronauts aboard an aircraft carrier.

The Soviets, however, had no such option. Their Navy, at least in the early 1970s when the American shuttle was being conceived, was little more than a coastal defense force, hardly fit for the task of recovering spacemen at sea. Also, there was a security threat involved. Suppose the Soviets brought a capsule down in the sea and the U.S. Navy reached it first?

The Russians were left, then, with only one course of action—to bring spacecraft back to land within the boundaries of the U.S.S.R. That meant a thumpdown on land instead of a splashdown at sea, and the Soviets soon learned that recoveries on land could be risky indeed.

Voskhod II, launched on March 18, 1965, developed

trouble with its autopilot, which was supposed to control re-entry. The spaceship's rockets did not fire as required to slow the craft for its return to earth. So cosmonauts Alexei Leonov and Pavel Belyayev fired the rockets by manual control, and the Voskhod started down through the atmosphere. The capsule came down not in Kazakhstan, as anticipated, but in the Ural Mountains, 3,000 kilometers north of the planned landing site. To make matters worse, it landed in a forest and was soon surrounded by hungry wolves that tried to enter the capsule. Fortunately, Leonov and Belyayev managed to keep the animals outside the hatch, and were rescued the following morning by a ski patrol. The spacemen traveled on skis to a nearby clearing where a helicopter was waiting to carry them to warmth and safety.

The first sign of a Soviet shuttle project was seen by Western eyes in the late 1970s. A reconnaissance satellite passing over the U.S.S.R. snapped a picture of a delta-winged craft with a truncated stern and a configuration that strongly suggested it was built for re-entry from space.

There it was—the long-awaited Soviet shuttle.

The first published picture, an artist's visualization, appeared in *Aviation Week* early in 1978. The accompanying story said the Soviet shuttle was undergoing atmospheric flight tests. (The U.S. carried out similar tests with its shuttle in 1977, releasing the orbiter from a specially modified Boeing 747 jetliner over the Southwestern desert.)

The cat being out of the bag, it was only a matter of months before the Soviets admitted they were building their own space shuttle. In October, 1978, Radio Moscow broadcast a feature about the new Russian spacecraft.

Though vague in detail, the report said the shuttle would have three rocket motors and, with its expendable booster rocket, would be about 67 meters long. The news release said the spaceship would have a diameter of about 6.5 meters "with fuel containers." If that figure stood for the wingspan of the Soviet shuttle, then the spacecraft would probably be a shade larger than the old U.S. Air Force X-20, Dyna-Soar.

Indeed, the Soviet shuttle sounds so much like an updated version of the Dyna-Soar that the Russian craft has been nicknamed the "Dynasoarski." The Soviets refer to it as the *kosmolyot*—spaceplane.

It is hard for Western observers to reconstruct the history of the *kosmolyot* project, but on the whole it appears to have been a Johnny-come-lately attempt to match the American effort.

Something like the *kosmolyot* was on the minds of Soviet designers around 1970, twenty-three years after Joseph Stalin sent his son, along with Serov and Tokadi, on their mission to kidnap Eugen Sänger for work on an antipodal space bomber. As the seventies began, the Russians were watching the American shuttle's development closely. Undoubtedly they studied the various designs shown in U.S. aerospace trade journals, and some of the early Soviet shuttle schemes were little more than reincarnations of Walter Dornberger's twenty-year-old Project Bomi: a small orbiter carried into space, piggyback, by a larger flyback booster stage. That configuration was considered briefly for the American shuttle, but discarded.

A few fuzzy details of the Soviet plans seeped into the West through conversations with Russian aerospace engineers at international conferences. The Russians' remarks indicated that they were working toward either a delta-wing design like that of the X-20, or a "lifting body" form —a wingless craft that would substitute its underbelly for the wing, to generate lift on re-entry.*

On the whole, however, the Russians were secretive about their shuttle blueprints, and they probably learned far more from their Western counterparts than U.S. and European designers learned from the Soviets. When the Soviets did release what they said were designs for their shuttle, the illustrations turned out to be little more than older American models, doctored a bit with an airbrush and adorned with large red stars. Some of the Russian pictures bordered on the preposterous. One artist's conception circulated widely in the West showed a sleek Russian rocket plane equipped with a ramjet—an air-breathing engine—for use in airless space!

Officially, the Soviets would neither quite confirm nor deny the existence of their space-shuttle program. Vladimir Shalatov said in 1973 that the U.S.S.R. had "no . . . space-shuttle program under way." Yet he said at a later time that "transport ships" would be needed to shuttle cosmonauts and supplies between earth and space stations in orbit. He hinted that the Soviets were planning a system like the American Dyna-Soar: a piloted orbiter put into space by some kind of unmanned booster.

By 1976, Shalatov was quoted as saying the Russians had "perfected" a shuttle which would take off horizontally from an airfield. But that same year an enigmatic series of test flights began—tests that suggested the Soviets were working instead on a shuttle more along the lines of the rocket-launched X-20.

* One of these rotund fliers, reminiscent of water beetles in shape, was seen in the opening sequences of the U.S. television series *The Six Million Dollar Man.*

There were three tests in the series: Cosmos 881/882, December 15, 1976; Cosmos 997/998, March 30, 1978; and Cosmos 1100/1101, May 22, 1979.

All three followed the same pattern. They were double missions, involving two spacecraft launched on a single booster. All were one-orbit flights, no more. Each time, the orbit was inclined about 52 degrees to the equator. And every launch took place at night, within a couple of hours of midnight Greenwich Mean Time. This peculiar set of circumstances was ideal for test-flying a winged re-entry vehicle.

The launch time puts the spacecraft back over the southwestern U.S.S.R. just before dawn. And with only a slight course correction—an easy thing for a piloted, winged re-entry vehicle to accomplish—the ship sails directly over Tyuratam in the last few minutes before sunrise.

The conditions are perfect for observing re-entry from the ground. As the Soviet spacecraft drops through the atmosphere, a glowing shroud of ionized particles surrounds it. By studying the fiery plume, Soviet aerodynamics experts can tell how the re-entry module is behaving on its tortuous glide through the upper air.

Shortly thereafter, the pilot of the spacecraft sees dawn breaking over the Aral Sea and Lake Baikal, just ahead. On the horizon, he can see the peaks of the Himalayas shining in the morning sun.

By this time, the spacecraft is soaring through the lower, denser layers of the atmosphere, on its way down for a landing in the Kazakhstan desert just south of Lake Balkash. Nose up and belly forward, the Soviet shuttle makes its approach for touchdown. Doors in the underside open. The landing gear extends. And about an hour and a half after lift-off from Tyuratam, the shuttle rolls to a stop on the desert pavement, trailing a cloud of dust behind it. Mission accomplished: the *kosmolyot* is home.

Eventually the Soviet shuttle—and there is no longer much doubt in the West that a *kosmolyot* is what the Russians tested on these three Cosmos missions—will touch down at Tyuratam. The Soviets have been building a long runway for the spaceship there, similar to the landing facilities for the American shuttle at Cape Canaveral, Florida. Then the craft will be refurbished and sent up again.

But what will the *kosmolyot* be doing in space? If the published descriptions of it are accurate, it lacks the cargo capacity of the American shuttle, which suggests that its primary function will not be to lift large payloads into orbit.

Reconnaissance is one possible job for the Soviet shuttle. A *kosmolyot* could serve as a satellite interceptor, though such work is unlikely, for reasons mentioned earlier. Could the *kosmolyot* be a space tanker, designed to haul fuel into orbit for satellites, perhaps including particle-beam projectors? That role has been suggested.*

Most likely, however, the Soviet rocket plane will be just what Shalatov said it would be—a ferry for use between the earth and manned space stations such as the currently orbiting Salyut 6. The Salyuts are part of a Soviet military man-in-space program similar to that

---

* The British journal *Spaceflight*, in a 1980 article on these Cosmos tests, states that many mysteries surround the missions. Why, for example, were *two* satellites sent up on each flight? If one was a *kosmolyot,* then what was the other? No one in the West seems to know, and the Russians are unlikely to tell.

which the U.S. Air Force had in mind for its star-crossed Manned Orbiting Laboratory; and the Salyuts may be only the first step toward a massive Soviet presence in the heavens.

TWELVE

Riddle in Space

WINSTON CHURCHILL once confessed his inability to fathom the intentions of Russia's leaders. "I cannot forecast to you the action of Russia," said Churchill. "It is a riddle wrapped in a mystery inside an enigma."

Russia's plans for the future of its military space program are likewise a mystery to Western observers today. But present trends seem to indicate that the Russians have in mind an accelerated man-in-space program in the decades ahead—a development with potentially grim implications for the West.

The Soviets, like the Americans, had long wanted a manned space station for reconnaissance and other military duties. So about the time the United States was preparing to send up its Skylab, the U.S.S.R. was getting ready to launch its Salyut. Early Salyut designs looked much like NASA's old MOSS, from the 1960s. The Salyut is a series of three cylinders that increase in diameter from the front of the station to the rear. Solar panels extend like wings from the station's flanks. The station has docking ports for the manned Soyuz spacecraft that carry crews to and from Salyut, and for the unmanned robot supply vessels that haul provisions up from earth during Salyut missions.

Early in its history, Salyut split into two different projects. One was a civilian program aimed mainly at scientific research without direct military application. The other was a military Salyut. Each has a clearly defined set of characteristics.

Civilian Salyuts fly high, at about 350 kilometers altitude, to help with astronomical studies and also to minimize air friction. The commander on those non-military missions is a soldier, but the flight engineer is a civilian. A non-military Salyut sends telemetry (transmissions of data) back to earth on a frequency of about 15 megahertz (MHz), or million cycles per second.

The interesting thing about the civilian Salyut missions is the considerable publicity given them by the Soviets. The Russians tell the world virtually everything about the missions, from the cosmonauts' taste in food to their daily schedule of meals. Photos of the crew at work in the station are released to the world press, and sometimes television pictures from the station are broadcast back to earth for showing on Soviet home screens. These space extravaganzas are milked for all the publicity they are worth, as was done on the U.S. civilian space missions.

The military Salyuts are treated quite differently. No publicity surrounds these flights. No pictures are released, as a rule, and the press is not informed of the mission's purpose. The crews' jobs are usually described as "routine." The military Salyuts fly much lower than the civilian ones—about 265 kilometers—and carry high-resolution photographic equipment in place of the astronomical telescopes and other scientific gear hauled aboard the non-military Salyuts. The low orbit is excellent for reconnaissance, but from time to time the station must be "kicked" into a higher orbit as atmospheric friction slows the Salyut and brings it near re-entry. Crews on these missions are composed entirely of military officers, who communicate with the ground on a frequency normally reserved for armed forces transmissions: 19:944 MHz. Unlike the civilian Salyuts, the military version is evidently equipped to serve as a manned *or* unmanned spacecraft.

Salyut had an inauspicious beginning. Though the first station was launched successfully in April, 1971, the crew sent up to it had problems docking with the station and was forced to return to earth.

The second mission ended in death for three cosmonauts. The crew boarded the station successfully in June, 1971, and stayed on board for almost a month. On their return to earth, however, a malfunction on their space capsule depressurized it and the spacemen died for lack of air before their craft touched down.

Then, in the summer of 1972, the Russians announced that a new Salyut would be sent into orbit shortly. Possibly because of a booster gone haywire, the launch failed.

The Soviets released no information about this launch, and even the date of it was uncertain. Most Western sources indicate that it took place on July 30, 1972.

All the while Russia had been trying for a successful Salyut mission, the Americans had been progressing smoothly and swiftly toward their first Skylab flight. Moscow was under pressure because Salyut had been intended, in part, to upstage the American manned orbital laboratory, and thus far had achieved nothing but a string of disasters.

The Soviets had high hopes, then, for Salyut 2, which was launched on April 3, 1973. Several days after entering orbit the spacecraft disintegrated. Western space observers believe one of two mishaps may have doomed Salyut 2.

One theory is that the uppermost stage of the heavy Proton booster rocket blew up and damaged the Salyut severely. If this had happened, however, the spaceship would probably have been unresponsive to signals from the ground; and several times the Salyut carried out orbital correction maneuvers on command from mission control. It seems more probable that something on the craft itself exploded, perhaps a fuel tank, and destabilized it. Tumbling madly, the Salyut would have quickly reduced itself to a hurtling cloud of metal debris—which is exactly what showed up on U.S. deep-space radars where Salyut 2 should have been.

Soviet space officials tried their best to save face when it became apparent that Salyut 2 was a wretched failure. On April 18th, four days after the Salyut broke apart, the Russians denied they had ever intended it to be a manned station (in fact, a crew was waiting to fly up to Salyut 2 when the station was destroyed) and, by the end of April, were saying simply that the spacecraft had finished its mission. Later in 1973, an official publication called the disastrous flight "successful" and said that information from the mission would be useful in building new spaceships. Like the top brass in Joseph Heller's *Catch-22*, the Russians knew that failure can sometimes be hidden by calling it triumph.

Although Salyut failed to steal Skylab's thunder, the Russians persevered with the project. Under the cloak of the all-purpose Cosmos satellite series, the U.S.S.R. on May 11, 1973, launched a third Salyut vehicle. Understandably nervous after the horrors of the first few flights, the Soviets preferred not to identify this satellite publicly as a Salyut, in case something went wrong yet again—which it did.

The satellite—Cosmos 557—evidently did not answer commands from the ground after it attained orbit, indicating that something on board had malfunctioned and left the station totally useless.

Communications frequency and other data identified the satellite as a Salyut, and Western observers think it was probably a civilian model. Late in May, Cosmos 557 dropped back to earth and Soviet morale dropped as well.

Thus far the Salyut project had been an almost unqualified fiasco. Meanwhile, the Americans were scoring a propaganda triumph with their first Skylab, launched on May 14, 1973. It began to seem as if nothing would ever go right for Salyut.

On June 25, 1974, the jinx finally lifted, with the successful launching of Salyut 3. On July 3rd it was boarded by two cosmonauts, Pavel Popovich and Yuri Artyukhin. Both were professional soldiers (the former a colonel, the latter a lieutenant colonel), which indicated to the West that it would be a military mission.

Salyut 3's announced mission sounded innocent enough. Tass said the cosmonauts would be studying the landforms of the earth's surface, atmospheric phenomena, and how the human body responded to the environment of outer space. But the real purpose of Salyut 3 appears to have been military reconnaissance. Popovich and Artyukhin had on board a camera with a focal length of 10 meters—far more powerful than would be needed to take pictures of the earth for purely scientific reasons.

As *Space World* reported in an article on the flight, no photos taken by that supercamera were ever released. Nor were pictures of the ship's interior made public, save for some fuzzy and unrevealing TV images.

If the spacemen had ever intended to study the effects of extended spaceflight on the human organism, they had no time to make a thorough job of it. On July 18th, only fifteen days after boarding the Salyut, Popovich and Artyukhin climbed back into Soyuz, disengaged from the Salyut, and sped back to earth, film in hand.

Another military mission followed later that summer. On August 26, 1974, Lieutenant Colonel Gennadi Sarafanov and Colonel Lev Demin flew up to the Salyut in their Soyuz 15 spacecraft. Exactly what their assignments were, it is difficult to tell. "Scientific research" was the most Tass would say.

Suddenly the gremlins that had vexed the first four Salyuts returned. The Soyuz's automatic docking system started burning too much fuel, and the spaceship closed too quickly on the Salyut. Sarafanov and Demin tried to correct the trouble but failed, and the Soyuz had to return to earth prematurely. However, the mission was not a total loss. The reconnaissance camera on board the Salyut kept clicking away even in the spacemen's absence, and just over one month after Sarafanov and Demin's aborted flight, the Salyut ejected a recoverable film capsule and sent it back to earth.

The next Salyut station to be sent into space, in 1975, was the non-military Salyut 4. It followed the general pattern of peaceful Salyuts, including the relatively high orbit ill-suited for reconnaissance. Two crews visited this Salyut.

As Americans were preparing to celebrate their Bicentennial, in the early summer of 1976, the Russians were preparing Salyut 5, another military space platform, which was launched on June 22nd. Two weeks later, Soyuz 21 carried two cosmonauts, Colonel Boris Volynov and Lieutenant Colonel Vitali Zholobov, up to the space station.

Part of their business on Salyut was medical research. For the first two weeks in space, Volynov and Zholobov worked at a leisurely pace, making notes on their bodily condition and performing various tests and experiments. But the cosmonauts' workload increased sharply in the third week of their stay—and the following week, they spent a great deal of time communicating with the ground in code, an almost sure sign that military work was going on in the space station.

Most likely the cosmonauts had been told to abandon their research and fill in for a Soviet reconnaissance satellite, Cosmos 844, that failed while in orbit. So much work was loaded on the cosmonauts that by mid-August they were literally sick from their labors; and on August 23rd they returned to earth, so exhausted that they reportedly required a week to recover from their ordeal.

The jinx was still working overtime when Soyuz 23 went up on October 14th. The passengers, Vyacheslav Zudov and Valeri Rozhdestvenski, were both lieutenant colonels. They had to abort their mission when a radar malfunction made it impossible to dock safely with the Salyut. That misfortune was nothing, however, compared to what awaited them on the ground.

On re-entry, everything that could possibly go wrong did. They landed at night, in a snowstorm, with temperatures well below zero. And by sheer bad luck they splashed down in the only body of water in their recovery area: Lake Tengiz. As rescue forces tried to get them safely to land, the cosmonauts sat shivering in their capsule, unable to open the hatch and exit for fear of upsetting the Soyuz and drowning. It was dawn before frogmen were able to attach a line to the bobbing spacecraft, and a helicopter towed the Soyuz to shore.

The Soviets waited until an investigation of Soyuz 23 was finished before sending any more teams to the Salyut. On February 7, 1977, Soyuz 24 took off for the space station. This time the gremlins were merciful; the mission —which included telescopic observation of the earth and a few minor experiments in biology—went smoothly, and the astronauts returned to earth on February 24th.

That autumn, on September 29, 1977, the Salyut 6 station was put into orbit. The Russians have got more use out of this Salyut than all the others combined. Salyut 6 is much more sophisticated than earlier Salyut models, especially in its computer system, and cosmonauts on board the station have performed a wide variety of interesting experiments. They have tested, for example, a new kind of infrared telescope that uses supercold liquid gases to heighten its sensitivity to IR radiation.

One of the most significant features of Salyut 6 is its pair of docking ports rather than the single port of earlier Salyuts. Now a Salyut can receive two Soyuz spacecraft at once, or a Soyuz and an unmanned Progress supply vessel from earth. The Progress robot tanker/freighter is sent up to Salyut to bring supplies from earth and to ferry back film and other items from the space station to the ground.

Washington expected that another military Salyut would be launched soon after Salyut 6. Yet none has appeared, and to make matters even more puzzling, the Soviets announced in May, 1981, that they would be suspending manned space flights for as long as a year. Why this sudden hiatus in the Soviet manned space effort?

The best guess U.S. space experts can offer is that the Soviets have reached a plateau in their manned space program and are gathering strength for a quantum leap forward in the 1980s. If the double docking port on Salyut 6 is an indication, the Soviets are thinking in terms of future space stations capable of handling as many as eight Soyuzes at once. Since only four cosmonauts can fit inside a single Salyut at any given time, more Salyuts would be needed to accommodate the crews of the extra spacecraft. The result would be a large assemblage of Salyuts and other spacecraft hooked together in orbit—a "modular space station," as Western observers are calling it. New units could simply be "plugged in" as needed.

Perhaps to help pave the way for such stations, the Russians are said to be conducting intensive studies of "closed life support systems." At the Kirensky Institute in Siberia, scientists have been living in an isolated chamber with a totally self-contained ecological system. There is also evidence that the Russians are investigating ways to grow food in space. Various Salyut crews have tended small gardens in their spaceship, though the results were disappointing; the plants sprouted normally but then refused to produce seeds or fruit. This strange behavior has been explained as a consequence of zero gravity. It seems the absence of gravity affects the flow of fluids inside the plants, and consequently they had trouble ridding their cells of the waste products of metabolism. This problem could be overcome, of course, by using centrifugal force to provide a mock gravitational field, as von Braun suggested for the first manned space stations.

Experiments like these suggest that the Soviets are thinking of putting large numbers of men (and presumably women) into space for extended periods of time in the near future. Perhaps we are seeing, in these tests and plans, the beginnings of the first space colony.

Future Soviet space stations may resemble the wheel-like structure seen in Stanley Kubrick's movie *2001*. In 1980, *Aviation Week* reported that the Soviets were thinking of building a 100-ton station—bigger than Skylab—with a crew of twelve. To put the station into orbit, sometime in the mid-1980s, the Russians are thought to be building a rocket booster more powerful than the giant Saturn V that sent America's Apollo astronauts to the moon.

Such facilities in orbit could serve important military

functions. They would have room on board for a tremendous reconnaissance camera whose capabilities might put even America's Big Bird in the shade. If the Soviets ever decided to put laser or particle-beam weapons in orbit, a space station would serve as a useful "construction shack" while the device was being assembled. And if the Russians chose to make their station into a fuel depot for laser ASAT weapons, they could build an electrolysis plant capable of turning out liquid hydrogen and oxygen to fill the tanks of their orbiting arsenal.

Thus far only the United States and Russia have launched space stations successfully. Soon they may be joined by a third nation, whose rapid rise to military might has upset the international balance of power in orbit.

THIRTEEN

Dragon in the Sky

SHORTLY AFTER the 1971 military coup that installed him as President and dictator of Uganda, His Excellency General Field Marshal Dr. Idi Amin Dada decided that his nation was not going to be left behind in the international space race. He declared that Uganda was starting its own space program.

The Ugandan government announced that a carefully selected cadre of astronaut candidates was undergoing physical training on a specially built obstacle course furnished with old automobile tires. Amin's "space program" was good for a laugh, and yet it illustrated the unpleasant fact that there are more and more national space programs—some with military potential—all the time.

For more than a decade after the launch of Sputnik I and the first Explorer, the United States and the Soviet Union had outer space virtually to themselves. They had a commanding lead in space technology and controlled the big launch vehicles needed to orbit such heavy payloads as reconnaissance satellites and manned spacecraft.

Except for each other, the two superpowers faced no serious competition in space. But that era came to an end in the spring of 1970.

On April 24th, at 1:30 P.M. Greenwich Mean Time, a multistage liquid-fuel rocket took off from a launch site at Shuang-ch'eng-tze, in Inner Mongolia, about 1,400 kilometers due west of Peking. The rocket lifted into orbit a 173-kilogram satellite designated China 1 or Mao 1 by Western intelligence agencies.

China 1 went into an orbit with a period of 114 minutes, an apogee of about 2,400 kilometers, and a perigee of 439 kilometers, inclined 68.44 degrees to the equator. The Chinese satellite broadcast 60-second concerts consisting of a Cultural Revolution song titled "The East Is Red"—the Chinese equivalent of "America the Beautiful" —followed by a 5-second intermission and then 10 seconds of audio signals.

Publicly, American and European space experts seemed amused by China's "orbiting music box" and took pains to point out what a crude production it was in comparison to U.S. and Soviet satellites. One analyst at the Institute for Defense Analysis in Arlington, Virginia, described China I as primitive, and it *was* unsophisticated by Western standards.

China I had a markedly elliptical orbit, for one thing. This showed the Chinese lacked the expertise needed to place a spacecraft into a neat circular orbit, as the Russians and Americans had been doing for years.

In short, the public was told that China I was no cause for alarm—a far cry from the official reaction to the launch of Russia's first satellite thirteen years earlier. But China's satellite confirmed two disturbing developments that had been suspected and feared for years.

First, China now had the capability, or would have it soon, to challenge Russia and America in the field of space technology. China I's size proved that China could send up reconnaissance satellites similar to SAMOS in the near future. Even more chilling was the thought that the Chinese might place nuclear weapons in orbit; they had tested an atomic device six years earlier.

Second, China I's success demonstrated that the Chinese were at least on the way to developing ICBM capability. Not long hence, the Chinese were likely to have nuclear-tipped missiles capable of wiping out targets thousands of kilometers away, in European Russia and the United States. Moreover, Chairman Mao Tse-tung had a long record of bellicose rhetoric toward the rest of the world, and particularly toward the U.S. and the U.S.S.R. On numerous occasions, Mao had proclaimed China's mission to carry the banner of Socialism across the world by force of arms if necessary. Thus long-range missiles in Chinese possession seemed likely to pose a serious threat to world peace.

Distressing as it was, China's first satellite came as no

surprise to knowledgeable China-watchers abroad. Western analysts had been aware for more than a decade that the Chinese had an intense rocket research and development program under way, and since the early 1960s it had been assumed that the Chinese would send up a satellite by 1970 or earlier.

Ironically, it was the Soviets, who as China's traditional archenemy had the most to fear from the Chinese ascendancy in space, that had done more than anyone else to make it possible. In the 1950s, during a brief spell of warm feeling between Moscow and Peking, Soviet Premier Nikita Khrushchev had offered the Chinese access to Russian military rocket technology. The Soviets also made a present to Mao of some intermediate range missiles similar to those the Russians tried to station in Cuba in 1962. These rockets were probably comparable in size and range to the Soviet SS-4 Sandal, a one-stage liquid-fuel rocket approximately 23 meters long and 1.7 meters in diameter, with a range of 1,400 kilometers. (Sandal is the code name supplied by Western intelligence.)

The missiles Russia gave the Chinese were not capable of international range, because the Russians had no wish to hand China weapons that might be used against the Motherland at some future date if relations cooled. Needless to say, that policy turned out to be a wise one, because the Sino-Soviet friendship was brief.

China's leaders, recalling how the Soviets had made mere client states of other Asian nations with gifts of aid and equipment, were skeptical of bears bearing gifts. So, in the early 1960s, Peking began to slander the "revisionists" in the Kremlin and the Russians huffily withdrew their assistance.

The Chinese already had enough hardware and training to provide the nucleus of their own missile program, however, and they set to work improving the rockets the Russians had provided. In this effort, they had unexpected—and unintentional—help from the most unlikely person imaginable: the fiercely anti-Communist U.S. Senator Joseph McCarthy of Wisconsin.

In public speeches, McCarthy would hold up a sheet of paper and state ominously that it listed a given number of Communists known to be in the State Department or elsewhere. The exact number varied from one speech to another, and no one in the audience ever saw or heard the names on the roster. But the speeches had the desired effect. In short order, McCarthy whipped up anti-Red feeling to hurricane intensity in the United States. It became dangerous for a person to belong to any but the most innocuous organizations, and doubly dangerous to be a foreigner in America, especially if one hailed from a Communist nation such as China.

Among the many Chinese who left the U.S. during the McCarthy years was Dr. Tsien Hsueh-shen, of the Massachusetts Institute of Technology, sometimes called "the Chinese Wernher von Braun" for his role in building the Chinese rocket program. Tsien was born in Shanghai, and came to the United States on scholarship in 1935; he quickly established his reputation as a gifted rocket designer. After World War II, he participated in studies of captured V-2s and helped to develop the Army's WAC-Corporal high-altitude rocket. He also envisioned a winged rocket ship, similar in many ways to the modern NASA shuttle, that would have glided back from outer space. After leaving the U.S., Tsien taught briefly at the

University of Toronto before returning to China in 1955.

Some years after Tsien's repatriation, China was torn by the Cultural Revolution. Thousands of scientists, engineers, teachers, and other intellectuals were removed from their labs and classrooms, and forced to work as ordinary laborers in the fields. They were the lucky ones. Other "enemies of the Revolution," as they were called, were tortured and in some cases even murdered by squads of Red Guards, roaming bands of young thugs who used Mao's authority to institute a reign of terror. This callous waste of brainpower set Chinese science back several decades in a matter of months. Nonetheless, rocket research went on.

By 1966, the Chinese had missiles capable of carrying nuclear warheads 650 kilometers—roughly the distance from Boston to Baltimore. Such a rocket was used for a bomb test at Lop Nor that year, which alerted the West to China's advances in rocket technology, and indicated that Peking would soon have orbital launch capability, most likely by adding one or two extra stages to the liquid-fuel rockets provided by the Russians.

About this time, China's first satellite launch seemed so close that the U.S. intelligence community expected to see a Chinese space vehicle in orbit by 1968. In fact, it took two years more than that. The cause of the delay is uncertain, but there were probably mishaps early in the program. Every nation has had at least one failure before sending up a satellite successfully (the U.S. Vanguard disaster, for instance); and while little information is available on launch attempts prior to China 1, it is believed there were two, and possibly three, previous unsuccessful attempts to orbit a Chinese satellite.

The official Hsinhua News Agency hailed China 1 as a great victory for the thought of Chairman Mao, who was quoted as saying China intended to build and launch still more earth satellites. Mao made good on that pledge on March 3, 1971, when China 2 went into orbit.

Slightly larger than its predecessor, the second Chinese spacecraft was described as an experimental satellite with both military and scientific tasks to perform. The inclination of its orbit, 69.9 degrees, was suitable for a spy satellite, for it carried China 2 over most of the land area of the United States and the Soviet Union.

There was a hiatus of more than four years between China's second and third satellites. The Chinese made up for the delay in 1975, however, by sending up *three* spacecraft.

China 3 was launched on July 26, 1975, and was remarkable for two things. One was its weight. At 3.5 tons, China 3 was an order of magnitude more massive than any other payload the Chinese had orbited. Also, China frankly admitted the satellite had a military job. Hsinhua said China 3 was part of a drive to "prepare against war," meaning defense against its mighty northern neighbor.

Officials in Peking were quiet about what kind of instrumentation China 3 carried, but most likely it was a combination reconnaissance vehicle and weather observation satellite. It transmitted data back to earth on frequencies suitable for long-range military communications, and stayed in orbit for almost two months. China 3 was brought down on command over the Pacific on September 14, 1975, and burned up on re-entry.

Less than three months after China 3 plunged back to earth, a fourth Chinese satellite reached orbit. China 4 was about the same size as its predecessor, but appears to have been the prototype for a more sophisticated kind of spy satellite.

China 4 flew at a low altitude (173 to 483 kilometers) good for taking pictures, and returned a small capsule—presumably filled with film—to earth about a week after launch. *Spaceflight*, in a 1979 article about the Chinese space program, pointed out that the tilt of China 4's orbit, 63 degrees to the equator, may have been chosen to make film retrieval easier. At that inclination, the satellite would spend long periods over Chinese territory, the better to catch any module ejected from the spacecraft. (Unlike film containers from U.S. satellites, which are recovered in midair over the seas by military planes equipped to catch the capsules' parachutes, Chinese film capsules must land inside China's borders, there being no far-flung Chinese navy or air force to collect them elsewhere.)

China 5, another 3.5-tonner, was sent up on December 16, 1975, and was virtually a replay of China 4, except that no film capsule was returned. It is believed the fifth Chinese satellite carried TV cameras that transmitted their pictures back to earth electronically. Probably because there was no film ejected, the Chinese reverted to a 69-degree inclination on this flight.

When the next satellite, China 6, was launched on August 30, 1976, it seemed an anticlimax after the three giant satellites that preceded it. Only 250 kilograms in weight, China 6 appeared to contain a scientific payload much like that of China 2.

But the relatively unimpressive China 6 was followed on December 7th by a blockbuster: China 7, the biggest satellite yet (3.6 tons), and one with sobering implications for the future of China's space effort.

China 7's orbit, inclined only 59.5 degrees to the equator, was so like China 4's that Western observers expected a capsule to be returned to earth. The huge *2.5-ton* re-entry module, as heavy as an automobile, gave Western analysts reason to think China might be planning manned space missions in the near future, a development that would have seemed unthinkable only a few years earlier.

The next satellite, China 8, was even bigger than China 7, possibly weighing as much as five tons. It also sent back a capsule. China 8 was sent into orbit on January 6, 1978. One month later, *Aviation Week* reported China 8's achievements and noted that a working re-entry system like that which the Chinese were developing would be a great aid in any future manned space program.

The following year, the Chinese themselves confirmed that they were working toward a manned spaceflight capability when one of the leaders of China's space effort, Jen Hsi-min, was quoted in *Aviation Week* as saying China was preparing to send up astronauts. He added that China wished to build space stations in orbit, like the Soviet Salyut and the American Skylab.

Less than a year later, it was reported in *Aviation Week* that the Chinese had chosen several astronaut candidates and were thinking in terms of a space program to be carried out over the next several decades. They seemed to be planning for a second generation of spacemen after the present crop of astronauts had gone into orbit, for the article mentioned that future astronaut instructors were being trained at an unidentified location, along with the spaceflight candidates themselves.

According to the descriptions in *Aviation Week*, Chinese training for space travel parallels closely that of the early American and Soviet astronauts. A few of the techniques mentioned were:

1. Centrifuge tests, to duplicate the high g-force loads that astronauts must endure on takeoff and landing.

2. Zero-gravity training, using aircraft that fly up-and-down patterns producing about half a minute of weightlessness at a time.

3. Rides in a swinging gondola, to give the trainees practice in handling a wildly tumbling spaceship.

4. Vibration tests, in which the subject is secured to a chair and then shaken violently, as he probably will be during launching.

5. Impact training, which subjects trainees (strapped to specially padded chairs) to simulated hard landings. Such landings are hazardous but a near certainty for Chinese astronauts, because the Chinese will be unable to recover their spacecraft from soft splashdowns at sea. Like Russia's re-entry vehicles, China's will have to touch down on dry land, and Russian experience has shown that those landings can be rough indeed.

6. A spacecraft simulator, to train astronauts in navigation and other skills.

The Chinese are also said to be working on such aspects of manned space missions as food. They have reportedly devised a "moon cake" that can be eaten without scattering crumbs about the cabin in zero gravity.*

* Feeding astronauts adequately while they are in space has been almost as great a problem as getting them there. In the early days, spacecraft designers wishfully thought rations could be com-

Though the Chinese are usually uncommunicative about the military side of their space program, the recent thaw in Chinese-American relations has encouraged Peking to open some of China's space facilities to foreign visitors, who have come away impressed with what they saw.

In 1979, a group of visiting space-technology experts from the United States toured more than a dozen space and aircraft installations in China, including the factory that assembles China's new CSS-X-4, an ICBM with a range of more than 9,000 kilometers, which was tested recently in flights over the Pacific Ocean from Shuang-ch'eng-tze. The CSS-X-4, known also to Western intelligence as the CSL-2, is comparable in most respects to the U.S. Titan liquid-fuel missiles. The CSS-X-4 is thought capable of delivering a nuclear warhead of 5–50 megatons' yield (a five-megaton warhead would wipe out a city

---

pressed into tiny volumes as food pills, each about the size of a vitamin tablet and containing a considerable amount of protein, minerals, vitamins, and so forth. That plan failed, because the human digestive tract requires bulk—fiber—to function properly, and that element of diet is hard to condense past a certain point. Squeeze-tube food can be packaged easily but is generally unappealing, mushy in texture, and reminiscent of meals for infants. Food in any form adds weight to a spacecraft, and extra weight means a smaller payload and shorter missions. One desperate scheme put forward around 1960 was to build an edible spacecraft. In this plan, the astronauts' rations would have been dried and compressed into a hard substance much like pemmican, which would then be used to build parts of the spaceship's interior. (A similar process is used to package victuals for mountaineers.) As they flew, the spacemen would make meals out of their vehicle. This plan was discussed seriously for a while, but finally discarded.

the size of Boston), and an advanced version of the CSS-X-4, the FB-1, was used to launch China 3 through China 8. The Chinese reportedly plan to modify the FB-1 to a three-stage launch vehicle with a liquid hydrogen/oxygen upper stage like those used in U.S. Saturn moon rockets.

At another facility near Peking, the American visitors saw a test stand used to fire hydrogen/oxygen motors like those intended for the FB-1, and were questioned closely about the use of liquid gases as fuels. The U.S. delegation suspected the Chinese had suffered setbacks in their rocket program because of their inability to use the supercold liquids as propellants.

An American team viewed a satellite control center at Xi'an and pronounced it roughly equivalent to the U.S. Air Force satellite control facility in Sunnyvale, California. At the Tung Fang Scientific Instrument Plant in Peking, they studied a model of what their hosts said was a recently designed unmanned satellite. About a meter in diameter and solar-powered, it was designed, the Chinese explained, to remain in orbit for years without being recovered. Such long-lived spacecraft are further evidence that China has extensive plans for space, and the technology to realize those plans.

The technology used on Chinese space missions is largely designed and built by the Chinese themselves, in keeping with the late Mao's dictum "We must rely on our own resources." Foreign students of the Chinese space program have been impressed with its nearly total reliance on Chinese-built hardware, as well as by the high quality of the equipment.

At the Institute of Technical Physics in Shanghai, for example, American visitors have found a highly advanced program in earth resources technology, including research and development work on infrared sensing devices which could serve military as well as peaceful ends, as components of Chinese space systems. IR detectors are already used in U.S. and Soviet satellites to monitor military activity on the ground, and it would seem that the Chinese intend the same use for some of their sensors.

Though China's achievements in space are impressive, considering how late their space program began, the Chinese are still lagging in many aspects of space technology. Weight reduction is one serious problem they face with their spacecraft. Chinese equipment, as a rule, weighs considerably more than U.S. and Russian devices designed to do the same jobs.

For assistance, Peking is looking to the U.S. and Western Europe, which thus far have been eager to help. The American government has given its blessing to Chinese shopping trips for military and civilian aerospace hardware, and China is trying to obtain many different kinds of space-system components from American manufacturers. Intrigued by TV broadcast satellites that U.S. industry built for the Japanese, China has expressed an interest in acquiring a powerful two-channel broadcasting system for the Chinese educational television network; and under an informal agreement reached in 1978 between Peking and Washington, the U.S. government and private aerospace firms will assist the Chinese in building their satellite broadcast capability. The U.S. space shuttle will probably be used to send up Chinese satellites. Chinese representatives have also been talking with the West Germans about a joint venture for putting communications satellites in orbit.

Naturally, *any* Chinese presence in space worries the

Soviet Union, but the repressive character of Soviet society might make Chinese broadcast satellites even more of a threat to Moscow than orbiting H-bombs. The Soviet government owes much of its power to its ability to censor the news reaching the Soviet people. That power would be jeopardized if the Chinese could broadcast propaganda directly into the U.S.S.R. by satellite. Only a small ground receiver would be needed to pick up the transmissions. All at once, the Soviet citizen, who normally is fed a carefully edited version of world events, would have easy access to news, entertainment, and other programing from outside the borders of the U.S.S.R. Chinese broadcasts might prove a powerful subversive force.

With that awful prospect in mind, the Russians have been stalling international negotiations on limiting the spread of arms in outer space. They have hoped to talk the U.S. into accepting a treaty that would protect U.S. satellites against Soviet assault, but leave Russia free to knock out China's satellites in the event the Chinese aim broadcasts into the Soviet Union by satellite. American negotiators have refused to allow such a flexible pact. Thus China's entry into the international space race has delayed this much-needed treaty. (At arms control talks, the Soviets have insisted that the U.S. shuttle be classified as an ASAT weapon and, as such, be discontinued. This extreme stand is seen in Washington as merely a bargaining tactic that Moscow would probably abandon if the U.S. conceded Russia the right to attack Chinese satellites at will.)

These, remember, are just some of the touchy issues surrounding China's *civilian* space program. China's military space program is sure to play even worse havoc with the international balance of power. The Chinese military presence in space might inspire them to adventurism. Soviet successes in space, it has been suggested, were among the reasons Soviet Premier Khrushchev became reckless in his foreign policy in the early 1960s. Ought we to assume such incaution is beyond the Chinese leadership?

What China has accomplished in her space program, other countries may soon try as well. As of this writing, more than a dozen nations and international organizations have sent payloads into orbit. They include:

Australia
Canada
European Space Agency (ESA)
European Space Research Organization (ESRO)
Federal Republic of Germany (West Germany)
France
India
Indonesia
International Telecommunications Satellite Organization (ITSO)
Japan
North Atlantic Treaty Organization (NATO)
People's Republic of China
Spain
U.S.S.R.
United Kingdom
U.S.A.

Some of the countries—besides the United States, Russia, and China—are building military space systems. The United Kingdom has had its Skynet satellite communications system in service for more than a decade. Tied to a ground terminal at a Royal Air Force facility near Alton, Skynet can carry high-priority British military messages around the world.

One of Britain's former colonies, India, has not formally entered the military space race as yet, but is beefing up its lauch capability. In 1980, India made a success of its SLV-3 satellite launch vehicle, a multistage rocket that put into orbit a 35-kilogram satellite. The next step, says the *New Scientist,* will be India's Augmented Satellite Launch Vehicle (ASLV), which is just an SLV with two strap-on stages attached to the main body of the rocket. ASLV could orbit a 150-kilogram satellite. China's first satellite was only slightly larger, and a few years after its launch China was sending up prototypes of spy satellites and manned spacecraft. May the world expect similar progress from India?

New Delhi surely has a motive to develop its own reconnaissance systems in space, for India has no lack of enemies, real and potential, who might require observation from orbit someday. Moreover, the Indian space program does not suffer some of the handicaps that faced the Chinese. There is no Maoist obsession with self-reliance, so the Indians are free to pick up hardware and expertise wherever they can find it.

Possibly, then, India's ascendancy in space will be even swifter than China's. With orbital launch vehicles will come ICBM capability, and now that India has tested a nuclear bomb successfully, that thought must give some of her neighbors sleepless nights.

Nor should the world ignore another Asian land with a long history of militarism: Japan. In February, 1980, the Japanese showed that their rockets could put a 350-

kilogram payload in orbit, and a bigger rocket, with a 500-kilogram payload, is in the works. Spy satellites could fit within those weight limits. So might military communications satellites.

Japan's most ambitious project, however, is its own space shuttle. A smaller version of NASA's, the Japanese shuttle looks almost like a scale model of Columbia. About 15 meters long and 8 meters in wingspan—about a third again as big as the Dyna-Soar—the Japanese orbiter will weigh 10 tons and will be able to carry a crew of four into space along with half a ton of cargo.

Does Japan have military work in mind for its minishuttle? So far, the talk has been mainly of using the shuttle to advance Japan's technology. As a nation built on high-technology industries, Japan cannot afford to let other nations take a commanding lead in that field. A Japanese shuttle in orbit, equipped with on-board factories turning out ultra-pure silicon and other valuable products, would surely help to keep Japan's businessmen competitive with the rest of the world.

But Japan's space effort is not merely a business enterprise. It is a government venture, supervised by Japan's equivalent of NASA, the National Space Development Agency (NASDA), which in its official literature uses a maritime analogy to describe its shuttle's mission.

Outer space, NASDA says in a brochure on the Japanese spaceship, is presently like a "virgin land" in the age of sailing vessels. Big fleets of ships were needed to exploit those lands, and it would seem Japan has plans for fleets of its spaceships to sail the new ocean of outer space. The last time Japan's government built a large fleet and sent it to sea, one of the results was the attack on Pearl Harbor.

FOURTEEN

"The Province of Chance"

''AR-RIGHT!''

The voice belongs to a young boy standing in an entertainment center at a suburban shopping mall. The walls are lined with computerized video games. For fifty cents a game, one can play basketball, football, hockey—or war in space. Today the youngster is blasting spacecraft with make-believe lasers.

The spaceships appear on the screen before him as stylized blobs of light with antennae. They weave and dart about unpredictably, all the while firing laser blasts—

symbolized by tiny dots of phosphorescence—at the boy's space fleet. His goal is to knock out more of the enemy's vessels than they can of his.

When one of the foe's space cruisers blows up with a realistic boom, the boy grins and crows his pleasure. When one of his own ships is stricken, he swears. At last an automatic timer inside the machine calls a truce. The final tally appears in glowing numerals at one corner of the screen: enemy losses 15, defenders only 5. Eyes agleam with satisfaction, the young warrior digs into the pockets of his cutoffs for more change.

Before he is much older, that boy may be serving in his country's Space Force or its equivalent. He may be called on to fight actual battles in space, firing real bursts of laser light instead of their graphic symbols on a monitor.

New Scientist, in a 1981 column, called space warfare "good" because it might bring about a return to the kind of combat practiced at sea during the days of Nelson and John Paul Jones. Eighteenth-century naval combat was, all things considered, an eminently satisfactory way to settle differences among nations when diplomacy failed. The ships themselves were cheap, at least by modern standards, and they released nothing noxious into the water if they sank. Battles normally took place far out at sea, and the civilians at home were spared the horrors of this particular kind of war. Moreover, engagements at sea were generally decisive. One side won, the other lost, and that was that. There were no years-long stalemates on the battlefield, no endless seesawing campaigns where troops had to pay in blood for the same real estate time and again. If wars had to happen, shoot-outs on the seas were probably the best way to wage them.

War in space, we are told, would mean a revival of this

kind of conflict. The ships would fire coherent light instead of chain shot, and the setting would be low orbit instead of the high seas; but otherwise, would outer-space combat be analogous to naval fighting in the 1700s—a tournament waged far away from homes and hearths as the populations of the belligerent countries stood by on the sidelines, so to speak, and waited for the victor to be announced?

Unfortunately, it would not.

Any future battles in outer space are unlikely to be jousts between knights in shining spacecraft, or romantic incidents like the *Constitution*'s fight with the *Guerrière*. The militarization of space will not spare our world the devastation of nuclear blasts by carrying combat off the earth and into orbit. That halcyon view of the military buildup in space overlooks one important point about these space systems: they were built not to replace our armies and navies and air forces here on earth, but to supplement and assist them in any wars they carry out on the planet.

It seems unrealistic, then, to expect that orbiting lasers and ASAT missiles are going to usher in a new era of postnuclear warfare, and make the atomic and conventional arsenals of the world obsolete. Spacewar systems may well increase the danger of war breaking out here on earth.*

* The American public has been given an extremely misleading picture of what orbital combat will entail, in part through certain motion pictures and television shows that depict war in space as a grand and ennobling adventure. The evil Klingons versus the virtuous Federation on *Star Trek*, the Empire versus the Rebel Alliance in *Star Wars*, and any number of space patrols and alien hordes in lesser epics have all contributed to the widespread idea

Perhaps the greatest danger posed by the militarization of space is that of war by accident.

At any given time, several thousand satellites and other pieces of equipment—spent booster stages and the like—are circling the earth, most of them in low orbit. The space immediately above the atmosphere has begun to resemble an expressway at rush hour. It is not uncommon for satellites to miss each other by only a kilometer or two, and satellites crashing into each other may explain some of the mysterious incidents in which space vehicles simply vanish from the skies. One civilian TV satellite has been lost in space; it never entered its intended orbit, and no signals were heard from it to indicate where it might have gone. Collision with something else in space seems a reasonable explanation of this disappearance. Even a tiny fragment of metal striking a satellite at a relative velocity of a few kilometers per second would wreck the satellite, ripping through it like a Magnum slug through a tin can.

Now suppose that kind of mishap befell a military satellite—in the worst possible situation, during a time of international tension with all players in the spacewar game braced for attacks on their spacecraft. The culpable fragment might be invisible from the ground; even something as small and light as a paper clip could inflict massive damage on a satellite at high velocity. Unaware of the accident, a less than cautious leader might interpret it as

---

that space warfare will be thrilling, dramatic, and fun. At the last moment, the photon torpedo (or whatever) will hit the foe in a vital spot, and the cosmos will be made safe for Truth and Justice in a single climactic explosion. That is the Hollywood model of warfare in space. It bears very little resemblance to what is likely to take place if combat ever occurs outside our atmosphere.

a preconceived attack. Wars have begun over smaller incidents.

Or, disaster may strike a satellite from inside. The rupture of an oxygen tank on Apollo 13 is a case in point; the blast wrecked much of the moonship's service module and came close to killing the three astronauts on the mission. It took hours for NASA to trace the cause of the accident. If such a calamity should strike a military satellite, particularly a large and important one, would our defense officials—or Russia's or China's—necessarily look for a non-belligerent explanation? They might not take time for a cool, step-by-step analysis of the accident if the threat of attack seemed imminent. When armed forces face each other, patience and rationality are often in short supply. And the finger on a trigger or a button may be itchy.

Potentially horrendous accidents occur in space with surprising frequency. One familiar example is the well-publicized crash of a Soviet Cosmos ocean surveillance satellite in northern Canada in the first days of 1978. The satellite destabilized itself unexpectedly—possibly because of a sudden venting of fuel—and plunged down along a track that would have brought the Cosmos, loaded with hazardous nuclear fuel, to impact in the New York City area if the descent had been delayed by one more orbit. We may not be so lucky the next time an incident of this kind occurs. And in some future spell of East-West tension, a chance event like the Cosmos's crash might trigger war. The problem of telling hostile spacecraft from harmless ones has not yet been completely solved; if a Soviet satellite were speeding toward impact somewhere in North America, with only Moscow's word to prove it posed no military danger, would our reaction be to wait and see if the Kremlin told the truth? We might interpret the spacecraft's approach as an attack—and react accordingly.

At a time like this, much would of course depend on luck, good or bad. The best simulations may turn out to be totally useless when the real emergency arises, because probability deals an unexpected hand. "His Sacred Majesty, Chance, decides everything," said Voltaire. And Clausewitz said:

> War is the province of chance. In no other sphere of human activity has such a margin to be left to this intruder, because none is in such constant contact with it on every side. It increases the uncertainty of every circumstance and deranges the course of events.

Clausewitz probably never foresaw that one day weapons would cruise the skies far above. But his remarks on the force of chance are particularly chilling when one considers how much it may do to upset our carefully made plans for military actions in space. A misunderstood or vague order, an omitted hyphen in a computer program, a fuel line left capped by oversight—small but significant happenings like these have determined the outcome of battles and space missions. May we safely assume nothing of the kind will happen as the armed forces expand their domain out into space?

Of course not. Gremlins will no doubt vex future military spacecraft, just as they did bomber crews and their planes during the Second World War. The difference is that spaceborne gremlins might cause an accident or misunderstanding that could lead to atomic war on earth. It takes little imagination to see how some unforeseen foul-up in orbit, or here on the ground, could create the impression of a satellite under enemy fire: an act of war. Fail-safe systems and careful engineering will naturally reduce the chance; but chance remains, and adds one more increment to the risk of Armageddon.

So much for accident. What about the possibility of *deliberate aggression* in space?

Premeditated war in this new arena is more than just conceivable. There are circumstances in which a quick first strike might seem attractive to Moscow.

As noted earlier in this book, United States armed forces rely heavily on satellites to carry their communications. Depending on which estimates one reads, up to 75 percent of all military messages to and from U.S. Army, Navy, and Air Force units pass through satellite channels. Much the same situation exists in the Soviet Union. The Russians, however, have shrewdly maintained cables and other on-the-ground channels for their military communications, in case something happens to their satellites. America has phased out many of its backup systems for satellite communications. Our satellites are therefore vulnerable in a way that Moscow's are not: if we lose our satellites, U.S. armed forces might find themselves at a serious disadvantage—"blinded," "deafened," and unable to respond effectively to a sudden, well-coordinated attack by land, sea, or air.

Let us assume, however, that Russia avoids tangling with America. The Soviets might attack the spacecraft of their other rival—China. It must be a great temptation to launch an ASAT missile at one or more of China's satellites, now that Peking's ambitions in orbit have been made known so clearly. How long can that temptation be resisted? And how would the United States react to an ASAT assault on China's space systems? Would we step in to aid the Chinese, whose spaceflight capability the West has done so much to develop, or would we sit on the sidelines and allow the two Communist giants to slug it out in space without interference? Sometime in the not too distant future, that situation may force some difficult decisions from the President of the United States.

Finally, there is the risk of outer-space warfare set off by provocateurs—small nations with access to rocket technology and a motive to cause trouble among the superpowers. It has been suggested that even a tiny and undeveloped nation could, with simple equipment and a little cunning, wage ASAT warfare. Suppose a notorious troublemaker decided to try framing the Soviets for an attack on an American satellite. In theory, the task would be easy. Get a high-altitude sounding rocket capable of reaching the orbits of U.S. spy satellites. Pack the payload compartment with BB shot. (Gravel or paper clips would do as well.) Set a charge to go off at the desired altitude, and launch the rocket straight up. Detonate the charge. A cloud of pellets or whatever spreads out in the path of an oncoming U.S. satellite. At a relative velocity of several kilometers per second, the satellite hits the buckshot. The pellets tear through the hull, rupture the fuel tanks, cause explosive decompression, and generally demolish the spacecraft. The whole provocation could be carried out for only a few thousand dollars. Military satellites would be terribly vulnerable to this kind of assault. In the manner just described, a billion-dollar spacecraft might be destroyed at little cost.

Yet perhaps the vulnerability of our space systems

offers the best hope for limiting the spread of weapons into space. If a space platform costs billions to build and can be knocked apart like a house of cards for little more than pocket money by anyone with access to simple rocket technology, then the generals and admirals here and abroad may decide that wars in space are a quixotic enterprise after all. Otherwise we may become—like H. G. Wells's Martians—the victims of our own military adventure beyond our world.

Recommended Readings

The literature on the military uses of outer space is scattered through a wide variety of books and magazines. Among journals that regularly carry articles on military space projects are *Aviation Week, Astronautics and Aeronautics, Spaceflight,* and *Space World.*

Highly recommended for further reading are the following:

"The Beam Weapons Race." *Aviation Week,* October 2, 1978. A survey of the U.S.-Soviet race to develop particle-beam weapons, by one of the world's most authoritative aerospace publications.

Bell, T. "America's Other Space Program." *The Sciences,* December, 1979; reprinted in *Space World,* April, 1980.

Brief but informative overview of U.S. and Russian military activity in space and its detrimental effects on the civilian space program.

Caidin, M. *Wings Into Space: The History and Future of Manned Space Flight.* New York: Holt, Rinehart & Winston, 1964.

One of America's best-known aerospace writers, Caidin details in this short book the German military spaceship designs and the U.S. manned military spacecraft that followed them.

Daniloff, N. *The Kremlin and the Cosmos.* New York: Knopf, 1972.

Daniloff's book contains interesting historical anecdotes about the cosmonauts and about the civilian and military figures behind Russia's manned space program.

Douglas, J. "High-Energy Laser Weapons." *Science News,* July 3, 1976.

Slightly dated but still informative survey of prospects for battlefield use of lasers.

Easterbrook, G. "The Spruce Goose of Outer Space." *Washington Monthly,* April, 1980.

A highly critical appraisal of NASA's claims for the space shuttle, this article explores the overly optimistic and misleading future that the space agency has set forth for the project in its official literature; an excellent piece of investigative journalism.

Garwin, R. "Are We on the Verge of an Arms Race in Space?" *Bulletin of the Atomic Scientists,* May, 1981.

One of the most knowledgeable writers on the subject of modern weapons systems and arms control takes a critical look at proposed military space projects such as lasers and particle beams.

Gatland, K. *Manned Spacecraft.* 2nd edition. New York: Macmillan, 1976.

Pocket-sized illustrated guide to manned space vehicles, from the A-9 space bomber of World War II to Apollo and Soyuz.

Grey, J. *Enterprise.* New York: Morrow, 1979.

An entertaining history of NASA's space shuttle, this book touches on Air Force involvement in the project, and specifically how the shuttle's design was modified to fit Air Force requirements.

Gurney, G., and Gurney, C. *Cosmonauts in Orbit: The Story of the Soviet Manned Space Program.* New York: Franklin Watts, 1972.
A clearly written and abundantly illustrated study of Russia's civilian and military man-in-space effort.

Hecht, J. "Laser Weapons: A Status Report." *Analog,* October 1977.
A hardheaded analysis of prospects for offensive and defensive lasers.

Hochman, S., and Wong, S. *Satellite Spies: The Frightening Impact of a New Technology.* Indianapolis: Bobbs-Merrill, 1976.
Entertaining introduction to reconnaissance satellite technology; recommended for teenagers.

Klass, P. *Secret Sentries in Space.* New York: Random House, 1971.
The classic book on spy satellites, by an editor of *Aviation Week* magazine; details how reconnaissance satellites have helped to keep the peace between the U.S. and the Soviet Union.

Laurie, P. "Exploding the Beam Weapons Myth." *New Scientist,* April 26, 1979.
An interesting analysis of what would be needed to build and operate a particle-beam weapon, with reflections on the military budget for PBW work.

Ley, W. *Rockets, Missiles, and Space Travel.* New York: Viking, 1959.
Authoritative and lively history of rocket technology up to the early days of the Soviet-American space race. Contains details of Nazi space-bomber designs.

National Aeronautics and Space Administration. *Space Shuttle.* Washington, D.C.: U.S. Government Printing Office, 1976.
NASA's official statement of the shuttle's purpose, with its military duties relegated to two short references.

Oberg, J. *Red Star in Orbit.* New York: Random House, 1981.
A lucid history of the Soviet space program from its origins to the early 1980s, by one of America's prominent aerospace

writers. Oberg's book is especially valuable for its section on rocket designer Sergei Korolyov, who more than anyone was responsible for Russia's rapid ascendancy in the field of space travel.

Ordway, F., and Sharpe, M. *The Rocket Team.* New York: Crowell, 1979.
Valuable for its anecdotes of the early days of the U.S. space program, *The Rocket Team* describes the genesis of programs that would later become America's military space effort.

Parry, A. *Russia's Rockets and Missiles.* New York: Doubleday, 1960.
A highly readable account of the beginnings of Soviet military rocket technology. Foreword by Willy Ley.

Peebles, C. "The Manned Orbiting Laboratory" (Parts I and II). *Spaceflight,* April 4, 1980; June 6, 1980.
A revealing look at a U.S. military space project now all but forgotten.

Pirard, T. "Chinese 'Secrets' Orbiting the Earth." *Spaceflight,* October 1977.
A comprehensive overview of the Chinese space program from its origins to the date of writing.

Shelton, W. *Soviet Space Exploration: The First Decade.* New York: Washington Square Press, 1968.
A handy summary of Soviet progress in space during the first ten years after Sputnik. Chapter 15 deals specifically with Soviet military activities in space.

U.S. Air Force. *Soviet Aerospace Handbook.* Washington, D.C.: U.S. Government Printing Office, 1978.
This Air Force publication is valuable for its discussion of how the Soviets view aerospace power and America's use of it. Also, see the "fighter pilot" on page 172.

Zaehringer, A. *Soviet Space Technology.* New York: Harper, 1961.
A short introduction to Russian rocket boosters and satellites, with a section on how the Soviet military controls the U.S.S.R.'s space effort.

Selected Bibliography

"ABM Promise Seen in Space-Based Lasers." *Aviation Week,* October 9, 1979, p. 15.

"Accelerated Laser Weapons Program Urged." *Aviation Week,* August 4, 1980, p. 52.

"Accelerator Deemed Key to Beam Effort." *Aviation Week,* August 4, 1980, p. 60.

Adomites, P. "The Space Shuttle: What Can We Expect?" *Space World,* March 1981, pp. 4–6.

"AEC Military Space Funds Voted." *Aviation Week,* June 22, 1964, p. 26.

"AF Experience Points to Growing Role." *Missiles and Rockets,* March 26, 1962, pp. 80–83.

"The Air Force and Gemini." *Missiles and Rockets,* April 1, 1963, p. 46.

"Air Force Inertial Upper Stage Delayed Year; Cost Increases." *Aviation Week,* January 7, 1980, p. 19.

"Air Force Space Goals, Projects Defended by Research Staff Chief." *Aviation Week,* June 3, 1963, p. 32.

"Air Force Studies Space Trainer." *Missiles and Rockets,* September 3, 1962, p. 12.

"Air Force Tests Laser Aboard Airplane in Major Arms Advance." *Providence Journal-Bulletin,* January 17, 1981, p. A2.

"Air Force to Test Laser Beam Weapon." *Boston Herald American,* May 30, 1981, p. 2.

Aldridge, R. "Who Will Shoot First in Space?" *The Nation,* March 25, 1978, pp. 333–35.

Alexander, G. "After Success, Shuttle Facing Problems." *Boston Globe,* May 12, 1980, p. 4.

————. "USAF Aims at Military Space Supremacy." *Aviation Week,* October 2, 1961, p. 28.

Alibrando, A. "NASA Links Efforts to Defense to Broaden Appeal for Support." *Aviation Week,* October 2, 1963, p. 32.

————. "NASA Losing Ground Within Congress." *Aviation Week,* September 16, 1963, p. 28.

"America Loves Ya! Columbia Heroes Return in Triumph." *New York Post,* April 15, 1981, p. 1.

"America Returns to Space." *Boston Herald American,* April 5, 1981, p. A2.

"Anti-Satellite Laser Weapons Planned." *Aviation Week,* June 16, 1980, p. 243.

"Anti-Satellite Talk: A Case of [Deleted]." *Astronautics and Aeronautics,* February 1977, pp. 8–10.

"Antisatellite Weapon Design Advances." *Aviation Week,* June 16, 1960, p. 243.

"Apollo Use in Military Programs Proposed." *Aviation Week,* November 9, 1964, p. 27.

Armstrong, N. "Why We Need the Space Shuttle." *TV Guide,* March 14–20, 1981, pp. 29–30.

"Army Beam Weapons Program Moving to DARPA." *Aviation Week,* August 4, 1980, p. 51.

"Astronaut Candidates Training in China for Future Missions." *Aviation Week,* February 4, 1980, p. 52.

Baar, J. "Does the U.S. Have a Second Chance to Beat Russia?" *Missiles and Rockets,* December 5, 1960, pp. 12–15.

————. "Push to Speed Up Dyna-Soar." *Missiles and Rockets,* January 18, 1960, p. 11.

————, and Howard, W. "AF Attacks 'Secret' Navy Space Plan." *Missiles and Rockets,* January 18, 1960, p. 11.

Barber, P., and Spurr, R. "Spies in the Skies." *Atlas World Press Review,* August 1977, pp. 25–27.

"The Beam Weapons Race." *Aviation Week,* October 2, 1978, p. 9.

"Beginning a New Chapter in Space." *Boston Herald American,* April 10, 1981, p. 2.

Bekey, I., and Mayer, H. "1980–2000: Raising Our Sights for Advanced Space Systems." *Astronautics and Aeronautics,* July/August 1976, pp. 34–63.

Beller, W. "Soviet Efforts Are Closely Watched." *Missiles and Rockets,* September 11, 1961, p. 27.

Bierman, J. "U.S. Study Advises Lasers in Space." *Boston Globe,* March 1, 1981, p. 61.

Billman, K. "Radiation Energy Conversion in Space." *Astronautics and Aeronautics,* March 1979, pp. 18–26.

Blakeslee, A. "May Day in Space." *Omni,* March 1981, p. 42.

"Blue-Green Laser Eyed for Sub Communications." *Aviation Week,* July 28, 1980, p. 62.

"Boeing Studying Space Bomber Concepts." *Aviation Week,* April 10, 1961, pp. 26–27.

"Boon or Boondoggle?" *Boston Herald American,* April 15, 1981, pp. A1, A10.

Bowart, W., and Sutton, R. "Superwars: The Day of the Death Ray Is Here." *Gallery,* April 1981, pp. 37–38, 96, 98, 104.

Bradbury, R. "Beyond Eden." *Omni,* April 1980, pp. 89–90, 114, 116.

"Budget May Outstrip NASA's by 1970." *Missiles and Rockets,* March 26, 1962, pp. 38–39.

Burroughs, E. *Three Martian Novels*. New York: Dover, 1962.

Caidin, M. *Wings into Space: The History and Future of Manned Space Flight*. New York: Holt, Rinehart, & Winston, 1964.

Campbell, D. "Why the Soviets Put Uranium in Orbit." *New Scientist*, February 2, 1978, p. 268.

"Canadians Plan to Bill Soviets for Cosmos Debris Collection." *Aviation Week*, February 20, 1978, p. 24.

Carter, L. "Space: MOL to Give Military First Chance at Manned Flight; Soviet Reaction Unpredictable." *Science*, September 17, 1965, pp. 1357–59.

Cassutt, M. "The Military Salyuts." *Space World*, April 1979, pp. 18–25.

"Chinese Assemble FB-1 Booster from CSS-X-4 Missile." *Aviation Week*, August 11, 1980, pp. 18–20.

"China Beginning Manned Space Effort." *Aviation Week*, May 28, 1979, pp. 26–27.

"Chinese Formulate Requirements for Broadcast Satellite System." *Aviation Week*, March 31, 1980, p. 63.

"Chinese Get European Space Briefing." *Aviation Week*, September 26, 1977, p. 98.

"Chinese Missile, Booster Photographed." *Aviation Week*, February 5, 1979, p. 19.

"Chinese Orbit 381-lb. Satellite." *Aviation Week*, May 4, 1970, p. 24.

"Chinese Space Gains Hamper Antisatellite Limitations Treaty." *Aviation Week*, July 9, 1979, pp. 18–19.

"Chinese Technology Gains Found Extensive." *Aviation Week*, January 21, 1980, pp. 16–18.

"Chinese Tour U.S. Telecommunications Plants." *Aviation Week*, March 26, 1979, p. 53.

"Civilian Spacecraft Likely to Get Survivability Boost." *Aviation Week*, July 10, 1978, p. 16.

Clarke, A. *The Coming of the Space Age*. New York: Meredith Press, 1967.

Coleman, H. "Space Reconnaissance Expansion Urged." *Aviation Week*, November 20, 1978, pp. 57–58.

"Congress Warned of Eroding Space Capability." *Aviation Week*, April 6, 1981, p. 55.

Cooke, R. "The Pentagon Eyes the Shuttle." *Boston Globe,* May 17, 1981, p. A23.

———. "'We All Feel as Giants.'" *Boston Globe,* April 10, 1981, p. 2.

Cooper, J. "Self-defense in Outer Space." *Spaceflight,* September, 1962, pp. 164–68.

"Cosmonaut Training Chief Warns U.S. Against Military Use of Space Shuttle." *Boston Globe,* April 9, 1981, p. 10.

"Cosmonauts Use Tanker Supply in Refurbishing Salyut Vehicle." *Aviation Week,* March 19, 1979.

"Cosmos Debris Examined in Canada." *Aviation Week,* February 6, 1978, pp. 22–23.

"Cosmos 954: An Ugly Death." *Time,* February 6, 1978, pp. 28–29.

Coughlin, W. "Year Was Lost in Dyna-Soar Program." *Missiles and Rockets,* May 23, 1960, p. 14.

Cousins, N. "Security and Sophistication." *Saturday Review,* June 30, 1962, p. 14.

Covault, C. "Antisatellite Weapon Design Advances." *Aviation Week,* June 16, 1980, pp. 243–46.

———. "Carter Backs Shuttle Fund Rise." *Aviation Week,* November 19, 1979, pp. 16–18.

———. "Extensive Design Changes Mark Soyuz." *Aviation Week,* January 14, 1980, p. 57.

———. "Military Efforts in Space on Increase." *Aviation Week,* May 1, 1978, pp. 53–54.

———. "NASA Seeks Shuttle Capability Growth." *Aviation Week,* April 23, 1979, pp. 42–52.

———. "Progress Docking Extends Intensive Salyut Activities." *Aviation Week,* July 9, 1979, p. 20.

———. "Shuttle Concerns Force Action." *Aviation Week,* June 2, 1980, pp. 14–16.

———. "Shuttle Pivotal to Space Plans." *Aviation Week,* March 3, 1980, pp. 69–73.

———. "Shuttle Project Faces New Problems." *Aviation Week,* December 10, 1979, p. 20.

———. "Soviets Build Reusable Shuttle." *Aviation Week,* March 20, 1978, pp. 14–15.

————. "Soviets Developing Fly-back Launcher." *Aviation Week*, November 6, 1978, pp. 19–20.

————. "Soviets Plan Larger Space Assemblies." *Aviation Week*, October 9, 1978, p. 55.

————. "U.S. Pushes Antisatellite Effort." *Aviation Week*, July 17, 1978, pp. 14–15.

————. "U.S. Team Tours China Space Facilities." *Aviation Week*, June 25, 1979, pp. 77–82.

"DARPA Weighs Use of Blue-Green Laser for Communications with Submarines." *Aviation Week*, July 28, 1980, pp. 62–63.

David, H. "NASA May Get a Military Arm." *Missiles and Rockets*, August 5, 1963, p. 14.

————. "Role of Military in Space Could Be Expanded Under President Johnson." *Missiles and Rockets*, December 2, 1963, p. 16.

"A Day to Look Upward." *New York Times*, April 19, 1981, p. 14E.

"Defense Chief Terms Manned Shuttle Vital to Military Plans." *New York Times*, February 8, 1980, p. 14.

"Defense Dept. Experts Confirm Efficacy of Space-Based Lasers." *Aviation Week*, July 28, 1980, pp. 65–66.

"Densification Process Applied to Shuttle Tiles." *Aviation Week*, February 24, 1980, p. 22.

"Details of Titov's 17½-Orbit Flight." *Missiles and Rockets*, August 14, 1961, p. 15.

"Directed-Energy Effort Shifted." *Aviation Week*, August 4, 1980, pp. 44–47.

"DOD Balks Most Military Space Expansion Except in Reconnaissance." *Aviation Week*, March 11, 1963, p. 116.

"DOD Shuttle Mission Schedule Threatened." *Aviation Week*, August 4, 1980, p. 20.

"DOD Takes a Stand." *Missiles and Rockets*, October 22, 1962, p. 46.

Doris, M. "Spaced Out: The Liftoff and Beyond." *Boston Phoenix*, April 21, 1981, pp. 1, 17, 20.

Dornberger Sees Space as Military Arena." *Aviation Week*, September 18, 1961, pp. 57–59.

Dossey, J., and Trotti, G. "Counterpoint—A Lunar Colony." *Spaceflight*, July 1975, pp. 259–68.

"Doubts on Soviet Beam Work Dissolve." *Aviation Week*, July 28, 1980, p. 47.

Douglas, J. "High-Energy Laser Weapons." *Science News*, July 3, 1976, pp. 11–13.

Douglas, M. "Space Shuttle's Success Opens New Doors for Humankind." *Boston Herald American*, April 20, 1981, p. A15.

Douglas-Hamilton, D.; Kantrowitz, A.; and Reilly, D. "Laser-Assisted Propulsion Research." In *Radiation Energy Conversion in Space, Progress in Astronautics and Aeronautics*, 1978, pp. 271–78.

Dupuy, T. *The Evolution of Weapons and Warfare*. New York: Bobbs-Merrill, 1980.

Easterbrook, G. "The Spruce Goose of Outer Space." *Washington Monthly*, April 1980, pp. 32–48.

"Edwards Looms Large in Dyna-Soar." *Missiles and Rockets*, March 26, 1962, pp. 119–120.

"Emphasis on Military Systems to Grow." *Aviation Week*, March 3, 1980, p. 74.

"Excimer Laser Is an Excited Compound." *Aviation Week*, August 21, 1978, p. 39.

Ferguson, J. "USAF Studies U.S., Soviet Space Potential." *Aviation Week*, March 5, 1962, p. 75.

Fink, D. "Navy Feels Its MOL Experiment Valid on Military or NASA Flight." *Aviation Week*, December 28, 1964, p. 32.
———. "Station Holds Key to USAF's Man-in-space Mission." *Aviation Week*, March 16, 1964, p. 112.

Flitner, D. "Can Terrestrial Politics Be Kept Out of Space?" *Boston Globe*, April 18, 1981, p. 9.

"Flying Cathedrals." *Wall Street Journal*, April 15, 1981, p. 28.

Forbich, C. "The Soviet Space Shuttle Program." *Space World*, January 1981, pp. 4–8.

Forward, R. "Practical Anti-Gravity Still Far Off." *Missiles and Rockets*, September 11, 1961, pp. 28, 31.

"From Sputnik to Shuttle: Two Decades in Space." *U.S. News and World Report*, October 3, 1977, pp. 62–65.

"Gains by Soviet Reported in Test to Kill Satellites." *New York Times,* March 19, 1981, p. 1.

Galbraith, J. "The Economics of the Arms Race—and After." *Bulletin of the Atomic Scientists,* June/July 1981, pp. 13–16.

Gardner, T. "High-Priority Military Space Race Urged." *Aviation Week,* October 9, 1961, p. 30.

Garwin, R. "Are We on the Verge of an Arms Race in Space?" *Bulletin of the Atomic Scientists,* May 1981, pp. 48–53.

Gatland, K. "A Soviet Space Shuttle." *New Scientist,* June 8, 1978, pp. 676–78.

———. "A Thousand Cosmos Satellites." *New Scientist,* April 13, 1978, pp. 91–93.

Gillette, R. "Space Shuttle: A Giant Step for NASA and the Military?" *Science,* March 12, 1971, pp. 991–93.

Greenberg, D. "Military in Space: Air Force Seems to Have Won Argument for Expanded Program." *Science,* July 6, 1962, pp. 23–25.

———. "Soviet Space Feat: It Provides New Arguments for Military Role, Undercuts Lunar Landing Critics." *Science,* August 24, 1962, pp. 590–92.

Grey, J. *Enterprise.* New York: Morrow, 1979.

Griffiths, D. "Beam Weapon Impact Called Uncertain." *Aviation Week,* April 30, 1979, pp. 28–29.

Hager, D. "The Orbiting Junkyard." *Saturday Review,* September 5, 1970, pp. 44–45.

Hallion, R. "The Development of American Launch Vehicles Since 1945." In *Space Science Comes of Age,* edited by Hanle, P., and Chamberlain, V. Washington, D.C.: National Air and Space Museum, 1981.

Haugland, V. "Manned Space Flight." *Ordnance,* January/February 1963, pp. 430–33.

Hawkes, R. "Air Force Intensifies Space Campaign." *Missiles and Rockets,* September 24, 1962, pp. 16–17.

———. "Technical Details of X-20 Revealed." *Missiles and Rockets,* October 1, 1962, pp. 31–32.

———. "Welsh Says U.S. Will Build Manned Space Stations." *Missiles and Rockets,* September 16, 1963, pp. 16–17.

Hayes, I. "Britain's Military Satellite." *Spaceflight,* February 1978, pp. 67–68.

Hecht, J. "Laser Weapons: A Status Report." *Analog,* October 1977, pp. 68–83.

Henry, R., and Sloan, A. "Space Shuttle and Vandenberg Air Force Base." *Space World,* February 1977, pp. 29–36.

Hertzberg, A.; Sun, K.; and Jones, W. "Laser Aircraft." *Astronautics and Aeronautics,* March 1979, pp. 41–49.

"The High Ground." *Time,* August 24, 1962, p. 10.

"High-Intensity Electron Beams Pushed." *Aviation Week,* August 4, 1980, pp. 67–68.

"Hill Agency Urges U.S. to Develop Space Factories." *Washington Post,* February 14, 1980, p. A14.

Hooper, G. "Missions to Salyut 6." *Spaceflight,* March 1979, pp. 127–33.

"Hope of Increased DOD Space Funds Slim." *Aviation Week,* November 4, 1963, p. 83.

"Host of Craft in Orbit or Near Launch." *Missiles and Rockets,* March 26, 1962, pp. 41–44.

"House Votes Beam Research Authorization." *Aviation Week,* December 8, 1980, p. 36.

Houtman, M. "Albatross: The Soviet Space Shuttle." *Spaceview,* May/June 1976, pp. 24–31.

Hotz, R. "Military Role in Space." *Aviation Week,* October 7, 1974, p. 7.

———. "Military Space Problems." *Aviation Week,* November 30, 1959, p. 21.

———. "Military Space Role." *Aviation Week,* September 11, 1961, p. 17.

———. "Soviet Space Puzzle." *Aviation Week,* January 15, 1979, p. 7.

Hymoff, E. "Will the Next War Be Fought in Space?" *Popular Mechanics,* July 1977, pp. 47–51.

"India Rockets Ahead with Observation Satellites." *New Scientist,* June 11, 1981, p. 691.

"Indian Ambitions in Space Go Sky-High." *New Scientist,* January 22, 1981, p. 215.

"International Launch Vehicles." *Aviation Week,* March 3, 1980, pp. 108–09.

"Japan Gaining Maturity in Satellite Technology." *Aviation Week,* March 3, 1980, p. 92.

"Japan Pushes Space Launcher Development." *Aviation Week,* January 28, 1980, p. 51.

"Japan's Space Hopes Get Off the Ground." *New Scientist,* January 29, 1981, p. 291.

"Johnson Stress on Military Space Seen." *Aviation Week,* December 2, 1963, p. 26.

Joyce, C. "U.S. Report Calls for Research on Laser Weapons." *New Scientist,* January 8, 1981, pp. 52–53.

―――. "Weapons Come Under Fire at the A.A.A.S." *New Scientist,* January 8, 1981, p. 51.

Judge, J. "Fantastic Vehicles Will Be Realities." *Missiles and Rockets,* March 28, 1960, pp. 17–18, 22, 24.

Kanabayashi, M. "Japan Sets Mini-Entry in Space Race." *Wall Street Journal,* July 20, 1981, p. 21.

Kantrowitz, A. "Propulsion to Orbit by Ground-Based Lasers." *Astronautics and Aeronautics,* May 1972, p. 74.

―――. "The Relevance of Space." *Astronautics and Aeronautics,* March 1971, pp. 34–35.

Keith, D. "Soviet and U.S. Military Hardware: A Comparison as to Quality and Quantity." *Vital Speeches of the Day,* January 15, 1979, pp. 196–99.

Kennan, G. "A Brief Against Nuclear Overkill." *Boston Globe,* May 31, 1981, pp. A13, A16.

Kennedy, J. " 'If the Soviets Control Space . . . They Can Control Earth.' " *Missiles and Rockets,* October 10, 1960, pp. 12–13.

"Kennedy's Stand on Defense and Space." *Missiles and Rockets,* October 10, 1960, p. 50.

Killer Orb Threat to Essential Data Seen by Executive." *Space World,* January 1979, p. 20.

Kistiakowski, G., and Tsipis, K. "Outer Space Is No Place for the U.S.-Soviet Arms Race." *Boston Globe,* May 30, 1981, p. 15.

Klass, P. "Bambi ICBM Defense Concept Analyzed." *Aviation Week,* October 23, 1961, pp. 82–83, 87.

―――. "Chemical Laser Takes Weapons Lead." *Aviation Week,* August 28, 1978, p. 38.

————. "Fubini Urges Military Space Role Analysis." *Aviation Week,* September 2, 1963, p. 46.

————. "Laser Destroys Missile in Test." *Aviation Week,* August 7, 1978, pp. 14–16.

————. *Secret Sentries in Space.* New York: Random House, 1971.

————. "USAF Halts Saint Work, Shifts to Gemini." *Aviation Week,* December 10, 1962, p. 36.

————. "U.S. Reviews Satellite Telecast Policy." *Aviation Week,* January 23, 1978, p. 66.

Kolcum, E. "Russia Increases Military Space Tempo." *Aviation Week,* March 15, 1965, p. 113.

Kondracke, M. "The General Goes Zap." *New Republic,* July 2, 1977, pp. 23–25.

————. "The Space Shuttle and the Education Decline." *Wall Street Journal,* April 23, 1981, p. 27.

Large, A. "Dreams Will Face Reality in the Shuttle's Future." *Wall Street Journal,* April 23, 1981, p. 26.

————. "Troubled Space Shuttle Is Slated to Be Launched Early Next Month, with Much Riding on Success." *Wall Street Journal,* March 19, 1981, p. 56.

"Larger Shuttle Orbiter Fleet Urged." *Aviation Week,* February 5, 1979, p. 12.

"Laser Applications in Space Emphasized." *Aviation Week,* July 28, 1980, pp. 62–64.

"Laser, Particle Beam Combination Suggested." *Aviation Week,* August 7, 1978, p. 15.

"Laser to Test Against Missiles." *Aviation Week,* February 4, 1980, p. 25.

"Laser Use in Space Draws Special Focus." *Aviation Week,* August 21, 1978, p. 38.

"Laser Weapon Tested." *Boston Globe,* May 16, 1981, p. 24.

"Lasercom Experiment Package on STP to Include Receiver, Transmitter." *Aviation Week,* January 21, 1980, pp. 70–71.

Laurie, P. "Exploding the Beam Weapon Myth." *New Scientist,* April 26, 1979, pp. 248–50.

Lehman, M. *This High Man: The Life of Robert Goddard.* New York: Farrar, Straus & Company, 1963.

"Let's Demilitarize Space, Soviets Say." *Boston Herald American,* April 18, 1981, pp. A1, A4.

Lettenberg, M. "Don't Look Now But . . ." *Commonweal,* September 15, 1978, pp. 583–88.

Lewis, R. "Columbia Comes Home to a New Space Age." *New Scientist,* April 23, 1981, p. 208.

———. "Is NASA's Political Orbit Decaying?" *New Scientist,* February 15, 1979, p. 462.

———. "NASA's First 20 Years—and the Future." *New Scientist,* October 5, 1978, pp. 14–16.

———. "We Have a Super Vehicle Up There." *New Scientist,* April 16, 1981, pp. 140–41.

———, and Hewish, M. "To Space and Back." *New Scientist,* April 16, 1981, pp. 146–49.

Ley, W. "Cold War in Space." *Popular Science,* August 1966, pp. 41–45, 170.

———. *Rockets, Missiles, and Space Travel.* New York: Viking, 1957.

———. *Satellites, Rockets, and Outer Space.* New York: Signet, 1958.

Lyons, R. "Military Planners View the Shuttle as Way to Open Space for Warfare." *New York Times,* March 29, 1981, pp. 1, 34.

Lytle, S. "Soviet Space Laser Rumors Spark Defense Increase Bid." *Pittsburgh Press,* June 3, 1980, p. 6.

McKinnon, D. "If We Lose the Space Race, Russia May Bring Us All Down." *Boston Herald American,* March 15, 1981, p. A15.

"A Manned Military Space Laboratory." *Spaceflight,* May 1964, pp. 74–78.

Markoff, J. "NASA and the Pentagon: The Air Force Eyes a Star War." *The Nation,* January 7, 1978, pp. 16–18.

"MBB, Chinese Discuss TV Satellite." *Aviation Week,* March 31, 1980, p. 63.

McGuire, F. "Receptive Climate for Military Space Growing." *Missiles and Rockets,* June 24, 1963, p. 18.

———. "Space Accord Threatens X-20." *Missiles and Rockets,* January 28, 1963, pp. 14–15.

McPhee, J. *The Curve of Binding Energy.* New York: Farrar, Straus & Giroux, 1974.

Michener, J. "Space Exploration: Military and Non-military Advantages." *Vital Speeches of the Day,* July 15, 1979, pp. 578–80.

"Milestones," *Spaceflight,* September 1977, p. 305.

"Military and Commercial Use of Space Just Around the Bend." *Business Week,* August 20, 1960, pp. 26–27.

"Military Missions for X-20, Gemini Under Study." *Aviation Week,* March 25, 1963, p. 22.

"Military Needs May Aid Fifth Orbiter." *Aviation Week,* February 18, 1980, p. 24.

"Military Space Fight Rages." *Missiles and Rockets,* June 25, 1962, p. 13.

"Military Space Move Left to Russians." *Missiles and Rockets,* September 10, 1962, p. 16.

"Military Space Setback." *Missiles and Rockets,* January 28, 1963, p. 70.

Miller, B. "Aerospace, Military Laser Uses Explored." *Aviation Week,* April 22, 1963, p. 54.

"MOL Studies to Include Army, Navy Tasks." *Aviation Week,* December 16, 1963, p. 32.

"The Motto Is: Think Big, Think Dirty." *Time,* February 6, 1978, pp. 12–13.

Nahin, P. "The Laser BMD." *Analog,* October 1977, pp. 68–71, 84–107.

National Aeronautics and Space Administration. *Space Shuttle.* Washington, D.C.: U.S. Government Printing Office, 1976.

"NASA Studies Manned Space Station." *Aviation Week,* August 4, 1980, p. 19.

"Navy Schedules Laser Lethality Tests." *Aviation Week,* August 4, 1980, p. 55.

"New Colorado Operations Center to Provide System Focal Point." *Aviation Week,* June 16, 1980, pp. 250–51.

Newell, H. "Survival in the Space Age." *Aviation Week,* August 19, 1963, p. 21.

"No Future for Laser Weapons in Space." *New Scientist,* January 1, 1981, p. 3.

Normyle, W. "Gemini–4 to Carry Defense Experiments." *Aviation Week,* May 3, 1965, p. 65.

Oberg, J. "Plesetsk: Russia's Top Secret Military Space Center." *Space World,* March 1977, pp. 4–7.

———. *Red Star in Orbit.* New York: Random House, 1981.

———. "Salute to Salyut." *Analog,* December 1978, pp. 49–67.

———."The Why of Sputnik." *Space World,* December 1977, pp. 4–15.

"Orbiter Ferry to Space Center Delayed." *Aviation Week,* February 5, 1979, p. 13.

O'Toole, T. "NASA Gives Police Power to Captains of Spaceships." *Washington Post,* March 18, 1980, p. A8.

———. "NASA Set to Deflect Any Asteroid Threatening Earth." *Washington Post,* February 12, 1981, p. A7.

———. "Space Shuttle Money Pinch May Force Science to Take Back Seat." *Washington Post,* May 6, 1979, p. A3.

———. "Success Puts U.S. Ahead of Russians." *Washington Post,* April 15, 1981, pp. A1, A3.

———, and Balz, D. "Astronauts Fly to Perfect Landing." *Washington Post,* April 15, 1981, pp. A1, A4.

Ordway, F., and Sharpe, R. *The Rocket Team.* New York: Crowell, 1979.

Overbye, D. "Columbia, Gem of a New Ocean." *Discover,* June 1981, pp. 18–19.

Paine T. "The Next Century in Space: The Legacy of Our Apollo Mission to the Moon." *Vital Speeches of the Day,* September 15, 1979, pp. 734–36.

"Particle Beam Weapons?" *Technology Review,* August/September 1979, p. 82.

"Peace in the Heavens—Is It an Idle Hope?" *Newsweek,* October 8, 1962, pp. 72–80.

Peebles, C. "The Manned Orbiting Laboratory" (Parts I and II). *Spaceflight,* April 4, 1980, pp. 155–60; June 6, 1980, pp. 248–53.

———. "The Origins of the U.S. Space Shuttle." *Spaceflight,* November 1979, pp. 435–42.

"Pentagon Has Dibs on Space Shuttle." *New York Times,* February 10, 1980, p. E9.

"Pentagon Studying Laser Battle Stations in Space." *Aviation Week*, July 28, 1980, pp. 57–62.

Pirard, T. "Chinese 'Secrets' Orbiting the Earth." *Spaceflight*, October 1977, pp. 355–61.

Pournelle, J. "Lasers, Grazers, and Marxists." In *The Future at War*, vol. 1, edited by Bretnor, R., pp. 83–101. New York: Ace, 1979.

Powers, R. *Shuttle*. New York: Warner, 1980.

"Priorities Set for Anti-Satellite System." *Aviation Week*, September 3, 1979, p. 57.

Prochnau, B. "The Crew: A Veteran and a Space Rookie." *Boston Globe*, April 10, 1981, p. 2.

"Proper Focus, Firm Goals Urged for U.S. Military Space Program." *Aviation Week*, April 29, 1963, p. 27.

Pryor, R. "War in Space." *Science Digest*, May 1965, pp. 28–29.

Rao, B. "China: New Space Power, or Threat to Stability in Asia?" *Space World*, March 1981, pp. 19–20.

Recer, P. "Comeback in Space." *U.S. News and World Report*, February 23, 1981, pp. 58–60.

Ritchie, D. "Danger Out of the Blue." *Boston Evening Globe*, July 21, 1978, p. 19.

———. "A Defense Mission for the Shuttle?" *Baltimore Sun*, March 9, 1980, p. K4.

———. "Laser-Rattling in Outer Space." *Inquiry*, September 1, 1980, pp. 13–17.

Robertson, D. "Soviet Space Shuttle: What's It For?" *Astronomy*, April 1981, p. 26.

Robinson, C. "Air Force Emphasizes Laser Weapons." *Aviation Week*, October 30, 1978, pp. 51–55.

———. "Beam-Target Interaction Tested." *Aviation Week*, October 8, 1979, pp. 14–17.

———. "Key Beam Weapons Tests Slated." *Aviation Week*, October 9, 1978, pp. 42–53.

———. "Laser Technology Demonstration Proposed." *Aviation Week*, February 16, 1981, pp. 18–19.

———. "Space-Based Laser Battle Stations Seen." *Aviation Week*, December 8, 1980, pp. 36–40.

————. "Space-Based Systems Stressed." *Aviation Week*, March 3, 1980, pp. 25–28.

————. "U.S. Pushes Development of Beam Weapon." *Aviation Week*, October 2, 1978, pp. 14–22.

Ross, H. "Space Stations: To Be or Not to Be?" *Spaceflight*, October 1958, pp. 311–14.

"Russian Dyna-Soar Flying?" *Missiles and Rockets*, January 11, 1960, p. 13.

"SAINT Reoriented for Future." *Missiles and Rockets*, December 10, 1962, p. 14.

Schiefer, J. "Paul Warnke: Thinking the Unthinkable." *The Real Paper* (Cambridge, Mass.), February 19, 1981, pp. 7–8.

Schroeder, G. "How Russian Engineering Looked to a Captured German Scientist." *Aviation Week*, May 9, 1955, pp. 27, 30, 32, 34.

"Secret Meeting Surveys Progress." *Aviation Week*, October 9, 1978, p. 42.

"Secretary McNamara Outlines Defense Department Plans for Space." *Missiles and Rockets*, October 22, 1962, pp. 14–15.

Senate Committee on Aeronautical and Space Sciences. *Soviet Space Program, 1971–75*. Washington, D.C.: U.S. Government Printing Office, 1967.

Senter, R. "Banning Bombs in Orbit: Can the U.S. and the U.S.S.R. Get It Together?" *The New Republic*, October 3, 1963, pp. 16–18.

————. "The Military Moves into Space." *The New Republic*, September 11, 1965, pp. 11–13.

————. "That Attack from Space." *The New Republic*, May 4, 1963, pp. 17–19.

Shales, T. "Columbia, Gem of the Sky." *Washington Post*, April 15, 1981, pp. B1, B4.

"Shuttle Concerns Force Action." *Aviation Week*, June 2, 1980, p. 14.

"Shuttle Pivotal to Space Plans." *Aviation Week*, March 3, 1980, p. 69.

"Shuttle Slows Pace in Space Race." *New Scientist*, January 17, 1980, p. 139.

"Shuttle's Role in Space Race." *Boston Globe*, April 15, 1981, pp. 1, 10.

"Shuttle Support: High-Tech Problem Solving." *Langley Researcher,* August 1, 1980, pp. 1–2, 4.

Simpson, P. "The 'Star Wars' Defense Plan." *Boston Herald American,* February 15, 1981, p. A1.

Slay, A. "Space: The Air Force's Future Initiative." *Vital Speeches of the Day,* March 15, 1979, pp. 328–30.

Smith, B. "Air Force Seeks Shuttle Control Center." *Aviation Week,* November 12, 1979, pp. 21–22.

———. "Expanded Capabilities Seen for Military Space Systems." *Aviation Week,* March 12, 1979, p. 69.

———. "Ground-Based Electro-Optical Deep Space Surveillance System Passes Reviews." *Aviation Week,* August 27, 1979, pp. 48–53.

———. "Military Satellite Emphasis Increases." *Aviation Week,* January 29, 1979, pp. 164–67.

Smith, K. "The Balance of Power Has Shifted." *Boston Herald American,* April 18, 1981, pp. A1, A4.

Smith, R. "Back into Space with Columbia." *Science,* April 24, 1981, pp. 419–20.

———. "Military Plans for Shuttle Stir Concern." *Science,* May 1, 1981, pp. 520–21.

"Soviet Killer Satellite System Confirmed." *Aviation Week,* October 10, 1977, p. 19.

"Soviet Launches of More Military Salyuts Expected." *Aviation Week,* December 4, 1978, p. 17.

"Soviet Position on Space Shuttle Seen as Setback to Satellite Pact." *Washington Star,* June 1, 1979, p. A2.

"Soviets Build Directed Energy Weapon." *Aviation Week,* July 28, 1980, p. 47.

"Soviets Confirm Shuttle Vehicle Effort." *Aviation Week,* October 16, 1978, p. 25.

"Soviets Encrypting Space Links." *Aviation Week,* April 30, 1979, p. 26.

"Soviets Increasing Space Activities." *Aviation Week,* March 3, 1980, p. 83.

"Soviets Move Toward Plug-In Space Stations." *New Scientist,* January 19, 1978, p. 142.

"Soviets Now Put Stress on Space Shuttle." *Aviation Week,* March 11, 1974, pp. 20–21.

"Soviets Reveal Orbital Lab Space Plans." *Aviation Week*, April 9, 1979, p. 21.

"Soviets See Shuttle as Killer Satellite." *Aviation Week*, April 17, 1978, p. 17.

"Soviets Test Killer Spacecraft." *Aviation Week*, October 30, 1978, p. 17.

"Space-Age Arms Race." *Newsweek*, April 27, 1981, p. 39.

"Space Agency Ponders 15-Man Station." *Missiles and Rockets*, January 1, 1962, p. 11.

"Space Race? Russians Just Aren't in It." *New York Post*, April 15, 1981, p. 2.

"Space Shuttle Priorities Set for Payloads." *Aviation Week*, May 16, 1977, p. 13.

"Space Shuttle System Operations." *Space World*, January 1981, pp. 14–17.

"Space Surveillance Deemed Inadequate." *Aviation Week*, June 16, 1980, pp. 249–61.

"Space Treaty Proposal." *Aviation Week*, February 13, 1978, p. 7.

"Spaceborne Applications Emphasized." *Aviation Week*, July 28, 1980, p. 62.

"Spacecraft Survivability Boost Sought." *Aviation Week*, June 16, 1980, pp. 260–61.

Spurr, R., and Barber, S. "Spies in the Skies." *Atlas World Press Review*, August, 1977, pp. 25–27.

Stetson, J. "Air Force Secretary Emphasizes Space Operations." *Space World*, January 1979, p. 33.

Stone, I. "Apollo-Type, Six-Man Spacecraft Designed to Support 24-Man Orbital Space Station." *Aviation Week*, August 19, 1963, pp. 72–74.

"Stronger Directed Energy Effort Urged." *Aviation Week*, October 8, 1979, pp. 18–19.

"Success." *Boston Herald American*, April 15, 1981, p. A1.

"Supplemental Budget Passage Averts Added Shuttle Slippage." *Aviation Week*, July 7, 1980, p. 23.

Taylor, H. "NASA Starts Manned Station Studies." *Missiles and Rockets*, April 29, 1963, pp. 18–19.

"Teal Ruby Launch Delay Recommended." *Aviation Week,* September 17, 1979, p. 19.

"Technology Eyed to Defend ICBM's, Spacecraft." *Aviation Week,* July 28, 1980, p. 32.

"That Was One Fantastic Ride." *New York Times,* April 19, 1981, p. 1E.

"Thinking Matures on Military's Space Role." *Aviation Week,* July 22, 1963, p. 209.

"Titan III Seen Key to U.S. Security." *Missiles and Rockets,* April 15, 1963, p. 14.

"Tone and Pace." *Time,* August 31, 1962, pp. 13–14.

Torrey, L. "Doubts About Reports of Russian Beam Weapons." *New Scientist,* August 7, 1980, p. 435.

"Treaties Impose Uneven Constraints." *Aviation Week,* July 28, 1980, p. 66.

"Treaty Called Key Arms Control Measure." *Aviation Week,* December 19, 1966, p. 72.

Treleaven, M. "Missiles and Antimissiles." *America,* February 28, 1981, pp. 161–62.

Ulsamer, E. "Pervasive Importance of USAF's Space Mission." *Space World,* May 1977, pp. 12–16.

———. "Will the Soviets Wage War in Space?" *Space World,* May 1977, pp. 25–29.

"USAF Explores Strategic Space Plans." *Aviation Week,* September 28, 1959, p. 26.

"USAF Studies U.S., Soviet Space Potential." *Aviation Week,* March 5, 1962, p. 75.

"U.S. Effort Redirected to High-Energy Lasers." *Aviation Week,* July 28, 1980, p. 50.

U.S. House of Representatives. *The Next Ten Years in Space, 1959–1969.* Document No. 115, 86th Congress, 1st Session. Washington, D.C.: U.S. Government Printing Office, 1959.

"U.S. Nears Laser Weapons Decisions." *Aviation Week,* August 4, 1980, p. 48.

"U.S. Shuttle Has Brezhnev Worried." *Boston Herald American,* April 18, 1981, pp. A1, A4.

"U.S.S.R.: Keep Space 'Pure.'" *Boston Globe,* April 18, 1981, p. 4.

Vladimirov, L. *The Russian Space Bluff*. New York: Dial Press, 1973.

"War's Fourth Dimension." *Newsweek,* November 29, 1976, pp. 46–48.

"Weapons Laboratory Aids Beam Effort." *Aviation Week,* August 4, 1980, p. 56.

Weiss, R.; Pirri, A.; and Kemp, N. "Laser Propulsion." *Astronautics and Aeronautics,* March 1979, pp. 50–58.

"Where Are They Now?" *Newsweek,* October 8, 1962, p. 20.

"White Horse Concentrates on Neutral Particle Beam." *Aviation Week,* August 4, 1980, p. 63.

White, S., and Tokati, G. "Green Light for Soviet Space?" *New Scientist,* February 20, 1975, pp. 439–41.

"Why the Shuttle Matters." *The Economist,* April 18, 1981, pp. 12–13.

"Wild-Card Weapons." *Boston Globe,* March 22, 1981, p. A6.

Wilford, J. "Delay in Space Shuttle Facing Carter Review; Funds May Be Raised." *New York Times,* November 5, 1979, p. A1.

Wilks, W. "MOSS Plans Include Military." *Missiles and Rockets,* May 13, 1963, p. 14.

Williams, T. "Soviet Re-Entry Tests: A Winged Vehicle?" *Spaceflight,* May 1980, pp. 213–14.

Wilson, G. "GOP Campaign to Cite Aerospace Issues." *Aviation Week,* August 6, 1962, p. 32.

———. "House Unit Urges Separate USAF MOL." *Aviation Week,* June 7, 1965, p. 16.

———. "Sen. Cannon Presses for Space Deterrent." *Aviation Week,* August 20, 1962, p. 31.

———. "U.S. Is Formulating New Space Policy." *Aviation Week,* June 18, 1962, p. 26.

———. "U.S. to Increase Military Space Funds." *Aviation Week,* August 27, 1962, p. 34.

"X-20 Will Probe Piloted Lifting Re-Entry." *Aviation Week,* July 22, 1963, pp. 230–41.

Index

David Ritchie is a professional science writer who has written for numerous publications, including *Analog, Inquiry, Maine Life, The Boston Globe, Newsday, The New York Times Magazine* and *The Washington Post*. He is also the author of *The Ring of Fire*, a book about the circum-Pacific belt of seismic and volcanic activity. He has a B.A. in environmental sciences from the University of Virginia.